Manual of Water Supply Practices

M44

Distribution Valves:

Selection, Installation, Field Testing, and Maintenance

Third Edition

American Water Works Association

Manual of Water Supply Practices

M44

Distribution Valves:
Selection, Installation, Field Testing, and Maintenance

Third Edition

American Water Works Association

Manual of Water Supply Practices—M44, Third Edition

Distribution Valves: Selection, Installation, Field Testing, and Maintenance

Disclaimer
The authors, contributors, editors, and publisher do not assume responsibility for the validity of the content or any consequences of their use. In no event will AWWA be liable for direct, indirect, special, incidental, or consequential damages arising out of the use of information presented in this book. In particular, AWWA will not be responsible for any costs, including, but not limited to, those incurred as a result of lost revenue. In no event shall AWWA's liability exceed the amount paid for the purchase of this book.

If you find errors in this manual, please email books@awwa.org. Possible errata will be posted at www.awwa.org/resources-tools/resource-development-groups/manuals-program.aspx.

Project Manager/Senior Technical Editor: Melissa Valentine
Senior Manuals Specialist: Mindy Burke
Senior Production Editors: Cheryl Armstrong, Sandra Lankenau

Library of Congress Cataloging-in-Publication Data

Names: Morgan, Kenneth C. | Oberoi, Kanwal. | American Water Works Association.
Title: M44--distribution valves : selection, installation, field testing, and maintenance / by Kenneth C. Morgan, Kanwal Oberoi.
Other titles: Distribution valves
Description: Second edition. | Denver, CO : American Water Works Association, 2016.
Identifiers: LCCN 2015050635| ISBN 9781625760821 | ISBN 9781613003343 (eISBN)
Subjects: LCSH: Water-pipes--Valves. | Water--Distribution.
Classification: LCC TD491 .M6164 2016 | DDC 628.1/5--dc23 LC record available at http://lccn.loc.gov/2015050635

Printed in the United States of America

ISBN: 978-1-62576-082-1 eISBN: 978-1-61300-334-3

American Water Works Association

American Water Works Association
6666 West Quincy Avenue
Denver, CO 80235-3098
awwa.org

Contents

This page intentionally blank.

Figures

Tables

This page intentionally blank.

Preface

The importance of valves in the water distribution system cannot be overstated. Their effective selection, installation, operation, maintenance, repair, and replacement are vital to appropriately conveying an essential product to the customer. The primary focus of M44 is distribution valves; however, other valves essential to plant operations are also discussed. Specific to distribution systems, M44 is a guide for the selection, installation, field testing, and maintenance of water distribution valves.

This manual offers a brief history of valves and an overview of their theory and flow characteristics. It is a discussion of recommended practice rather than an AWWA standard calling for compliance with certain specifications. It provides guidance on generally available valve alternatives. Questions about specific situations or the applicability of a particular valve should be directed to the manufacturer or supplier.

Information contained in M44 is useful for operators, technicians, and engineers for all sizes of utilities, but it is particularly useful for small- and medium-sized organizations. The manual primarily discusses the more commonly used valves, but some of the "specialty" or "special use" valves are covered as well. The extent of the presentation of the specialty valves is only to introduce their availability and applications.

This manual provides general information to utilities and contractors on the proper installation and field testing of valves. Essential procedures related to operation and maintenance are also covered within this document. Additionally, procedures for determining valve criticality to assist an agency to establish, implement, and maintain an effective utilization of valves during their response to emergencies is provided as well.

Fire hydrants, although technically considered valves, are not covered in this manual. The reader is referred to AWWA Manual M17, *Installation, Field Testing, and Maintenance of Fire Hydrants.* This manual (M44) refers to AWWA standards for various types of valves. Copies of these standards may be purchased from the AWWA Bookstore.

Several manufacturers graciously provided valve illustrations and other documentation. AWWA does not endorse any manufacturer's products, and the names of the manufacturers have been removed from the material provided.

Metrication note: Valve sizes are listed in their current US designation; i.e., nominal pipe sizes in inches. To obtain an exact metric equivalent, use a conversion factor of 25.4 mm/in.

This page intentionally blank.

Acknowledgments

The AWWA Technical and Educational Council and the AWWA Distribution and Plant Operations Division gratefully acknowledge the contributions made by those volunteers who drafted, edited, and provided the significant and critical commentary essential to updating this manual. The Technical Review Board members dedicated many hours in the final stages of preparation of this edition to ensure the overall technical quality, consistency, and accuracy of the manual.

Technical Review Board Members

Kenneth C. Morgan, Chair of M44 subcommittee, Town of Gilbert, Gilbert, Ariz.
Kanwal Oberoi, Charleston Water System, Charleston, S.C.
Jonathan Peters, Greater Cincinnati Water Works, Cincinnati, Ohio
Greg Ramon, Little Rock Wastewater, Little Rock, Ark.
Charles Sparks, Los Angeles Department of Water and Power, Los Angeles, Calif.
Tim McCandless, American Water Works Association, Denver, Colo.

Contributors to the

Kenneth C. Morgan, Chair of M44 subcommittee, Town of Gilbert, Gilbert, Ariz.
Todd Alcott, Waterous Company, South St. Paul, Minn.
Larry Aulich, Mueller Company, Chattanooga, Tenn.
Dan Barr, Burgess & Niple Inc., Columbus, Ohio
Karl Borg, City of Hillsboro, Hillsboro, Ore.
David Bries, City of Montrose, Montrose, Colo.
Daniel Burczynski, Kennedy Valve, Elmira, N.Y.
John Cusack, Bryant Associates, Braintree, Mass.
Leo Fleury, Mueller Company, Chattanooga, Tenn.
Paul Gifford, Mueller Company, Chattanooga, Tenn.
Duane Gilles, Evansville Water Utility, Evansville, Ill.
Chad Harbour, Clow Valve Company, Oskaloosa, Iowa
Rhonda Harris, CH2M HILL OMBG, Dallas, Texas
Kirk Johnson, M&H Valve Company, Anniston, Ala.
Randy Looney, American AVK Company, Minden, Nev.
Kanwal Oberoi, Charleston Water System, Charleston, S.C.
Jonathan Peters, Greater Cincinnati Water Works, Cincinnati, Ohio
Greg Ramon, Little Rock Wastewater, Little Rock, Ark.
Raju Sandhu, Singer Valve, Surrey, British Columbia, Canada
Derek Scott, American Flow Control, Birmingham, Ala.
Andrew Slotterback, Burns & McDonnell, Kansas City, Mo.
Carl Smith, Val-Matic Valves & Manufacturing Corp., Elmhurst, Ill.
Ted Stolinski, Black & Veatch Corp., Kansas City, Mo.
Cliff Wilson, Wachs Water Services, Buffalo Grove, Ill.

This page intentionally blank.

Chapter 1

History and Design Considerations

A water distribution system is a piping network that delivers water at service pressure from the source to the end-user connection. Generally, the system originates at a treatment plant or pumping station and ends at a residential, industrial, or commercial service connection. The system may also terminate at a fire hydrant or connect to a subordinate system. Valves are a significant component of any water distribution system and are most commonly used for isolating a section of a flow line, controlling the flow, releasing air, and preventing backflow.

The topics in this introductory chapter include a brief historical review of valves and major design considerations, such as flow resistance, flow-control elements, and sealing mechanisms.

HISTORY

Valves have been a vital component of every water system from the water systems of Greece in 600 BC, which used flap valves to control water for showers, to the London Water Works and the first practical gate valve in the 1800s, to most modern water distribution systems used today. What is believed to be the earliest form of the mechanical valve, the bellows, is shown in hieroglyphs on the wall of an Egyptian tomb built more than 3,500 years ago. These wall paintings depict a bellows with bamboo piping used to smelt ore. Ever since, people have been refining and designing valves to fit different applications.

As water distribution systems became larger and more complex, particularly after the development of cast-iron pipe, designers met the challenges of water flow requirements with improved valve designs. However, many of the valve types currently used in water utility systems were developed in the 1800s and are still used with little change in their basic features. Nonetheless, hundreds of new designs and refinements are available to system designers. Figure 1-1 shows a historical timeline for valve development.

600 BC
Grecian illustrations show evidence of a flap-type valve used in showers for bathing.

400 BC
Plug-type valves are used by both the Greeks and Romans in water mains and supply pipes.

250 BC
Greeks develop a two-cylinder pump that employs a valve in each cylinder.

100 BC
Hero of Alexandria develops the first workable hinged valve. He advises that the seat facings should be smooth and polished to ensure a reasonably tight closure.

1400s
Leonardo da Vinci draws valve constructions, including a multiport check valve with four pairs of hinged gates.

1455
First cast-iron pipe is manufactured in Viegerland, Germany, for installation at Dillenburg Castle.

1600s
King Louis XIV of France orders a cast-iron distribution system to supply water to Versailles.

1700s
Pressure conduits are in general use.

1795
Joseph Bramah designs the first screw-thread valve stem. It is used to open and close sluice-type valves.

1800s
The designs of most commonly used valves are developed during this century. Solid wedge, taper seat valves are introduced in the United States, and most of the changes and refinements take place before 1900.

1839
James Nasmyth designs the first practical metal gate valve, which consists of a cast-iron body and bonnet, a threaded valve stem, and a solid tapered plug. It is immediately employed in England by the East London Water Works.

1875
S.J. Peet patents the double-disc, parallel-seat, bottom-wedge-type gate valve.

1900s
The butterfly, ball, and diaphragm valves are designed and their use becomes common.

Figure 1-1 A historical timeline for valve development

System designers have many choices when they select valves for use in the water distribution system. Making the right valve choice depends on many factors, including the size of the pipeline, the hydraulic pressures the valve will control, the material from which the valve is made, the material with which it will come in contact, what it will be used for, and cost. The following section describes basic design considerations when selecting a valve.

DESIGN CONSIDERATIONS

This section discusses important factors to consider when selecting valves. The first factor discussed is flow resistance, or friction, which causes head loss (loss of pressure) as the water flows through the system. The consideration of the loss of pressure through the system can directly impact the type and size of valve selected for a portion of a design project or for the entire design project. The other factors discussed are general valve elements involved in flow control, such as adjusting the flow and the sealing mechanism.

Flow Resistance

The term flow resistance refers to a fluid's resistance when moving through the piping system. Flow resistance plays an important role when designing a water distribution system. Flow resistance impacts flow pressure, pumping requirements, pipe size, and water volume, as well as the selection or type of valve needed for the specific application. A pressure drop caused by friction impacts the amount of water that can move through the pipe and valves. This drop is caused by fluid particles rubbing against each other and against the pipe. One can observe the effect of friction by connecting pressure gauges to a pipe system at both the beginning (upstream) and the end (downstream) of the system or to an isolated portion of the system; with gauges at the same elevation, the upstream gauge will show higher pressure than the downstream gauge.

Different pipe materials experience different levels of friction and so are said to have different friction factors. For example, a copper pipe is smoother than an iron pipe, and water flowing through the copper pipe will encounter less friction. In addition, for a given nominal diameter, the inside actual diameter may vary depending on pipe material. Flow is affected not only by the roughness of the pipe but also by the pipe diameter; friction increases as the diameter of the pipe decreases. The pressure drop associated with flow through a valve, i.e., the friction loss in the valve, depends on the type of valve.

Many equations allow distribution designers to calculate the pressure drop in a given system. System designers can calculate average flow resistance based on the turbulence of the flow, the roughness of the pipe, pipe diameter, valve and bend types, and other factors. Generally, the flow resistance of a valve is expressed in terms of the length of straight steel pipe that would have an equivalent resistance. For example, a 12-in. gate valve is rated at about a 13.2-ft (4-m) equivalent length of clean schedule 40 steel pipe, and a 12-in. globe valve is rated at about 340 ft (100 m). The relative resistance of PVC is about half that of steel.

The considerable complexity of the flow resistance of valves is beyond the scope of this manual. For more information, consult the sources listed in the additional readings at the end of the manual and manufacturers' technical data.

Flow-Control Elements

The many different types of valves have basic functions and elements that are similar. This section will discuss those basics. Figure 1-2 shows the basic valve elements.

Flow control. A valve controls flow through one of four basic closure methods:

- A disc or plug moves against or into an opening
- A flat, cylindrical, or spherical surface slides across an opening
- A disc or ellipse rotates across the diameter of a pipe or circular element
- A flexible material moves into a flow passage

The valve parts that control the flow element, or closure member, are the stem and the operating mechanism—a handwheel, lever, or key that attaches to the operating nut.

Stems. Most valves employ a threaded stem to move the flow element. Although there are exceptions, such as safety valves and check valves, most valves have stems that extend to the outside of the valve. A rotating stem provides movement for nonrising-stem gate valves, ball valves, rotating-disc gate valves, butterfly valves, and most plug valves. Quick-opening gate valves, rising-stem gate valves, globe and diaphragm valves, and outside-spring safety and relief valves operate with a stem that moves axially.

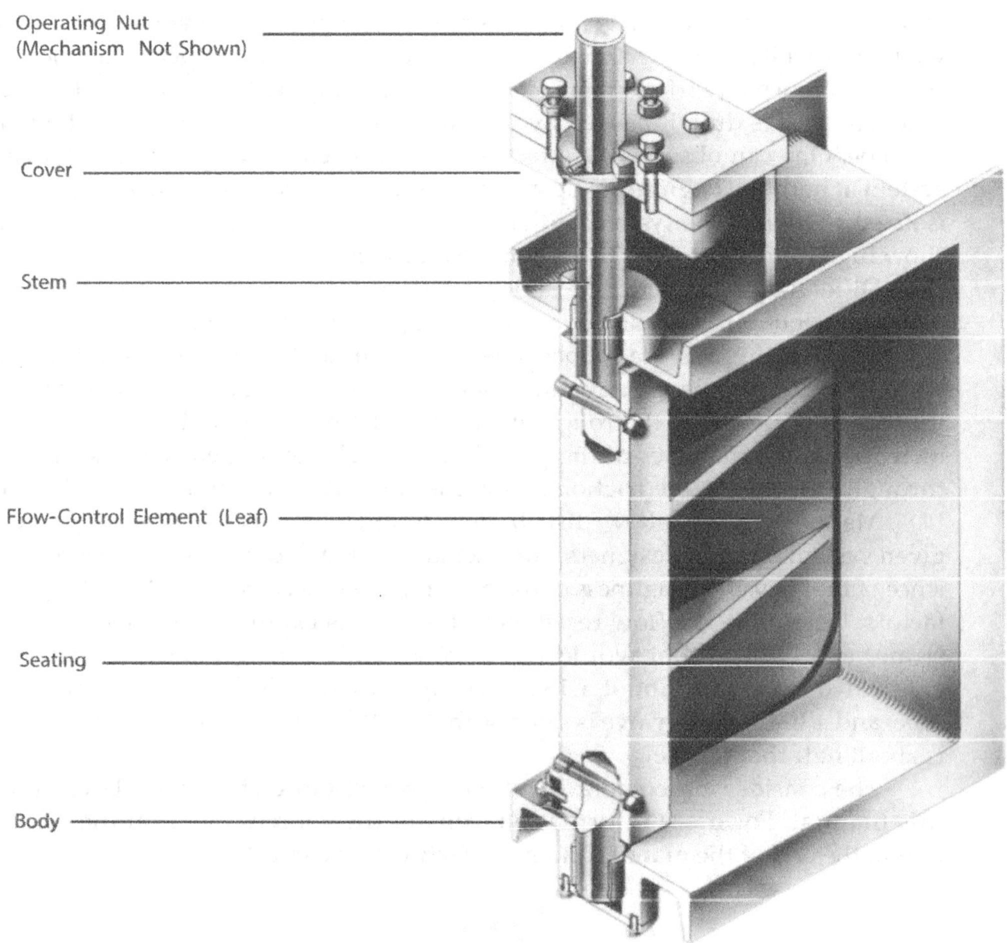

Figure 1-2 Valve elements (rectangular butterfly valve)

Some stems both rotate and move axially, such as globe and needle valves, lift-type cone or plug valves, and most pinch valves. The factors that influence stem design and selection include torque, thrust, valve size, and physical parameters.

Valve operation mechanisms. Most valves are operated with a handwheel (Figure 1-3), operating nut, or lever. Not every application lends itself to such a simple solution, so designers have devised a variety of methods for both manual and automatic valve operation. Special methods are usually needed when special conditions exist, such as:

- The valve must be operated remotely.
- The secure operation of the valve is critical.
- The valve is inoperable using normal methods.
- The size of the valve makes it impossible for one person to operate the valve.

The bury depth of valves may vary. Stem extensions or adjustable valve keys have been designed to operate most valves. Stem extensions usually consist of a steel rod and a coupling that attaches to the valve stem. If the extension must be very long to reach a valve, extra support is provided to keep the extension rigid and to prevent bending or breaking. Operators can use adjustable shafts or steel rods and universal joints to reach valves in difficult locations.

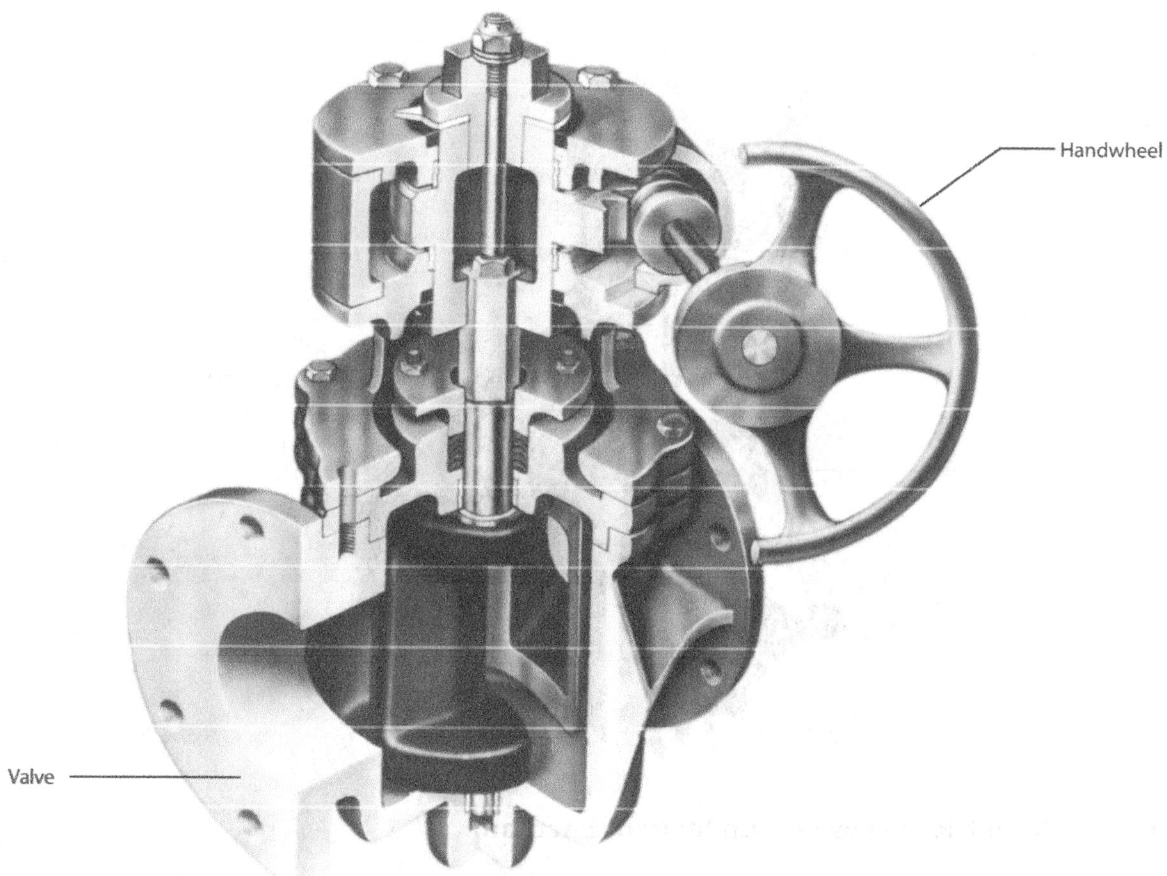

Figure 1-3 Eccentric plug valve with handwheel actuator

Floor stands, gear operators, and wheel operators give personnel a mechanical advantage in opening or closing valves that are inconvenient to reach or are large and difficult to operate. Position indicators may be installed on nonrising-stem valves to show how far open the valve is.

Accessories for automatic operation are also available. They can be used simply to open or close a valve or to throttle flows. Hydraulic or pneumatic actuators, which operate with a diaphragm or piston construction, are common. In a piston type, two chambers in a cylinder are isolated from each other by a piston. The valve stem is connected to the piston. As hydraulic fluid or air is pumped to one side or the other of the piston, the piston is forced back and forth inside the cylinder, which operates the valve. A typical hydraulic operator installation is shown in Figure 1-4. Electric motors are also used to operate valves.

Sealing Mechanisms

Valves are employed to control the flow of fluid through a piping system. Valve seatings are the portions of the valve that contact the valve body to form a seal that stops or diminishes the flow of liquid. Because they undergo wear during the sealing process, they will become less effective over time. Valve-sealing mechanisms used in water distribution systems are usually metal seatings or soft seatings. Another sealing mechanism involves using a sealant such as that used in a lubricated plug valve.

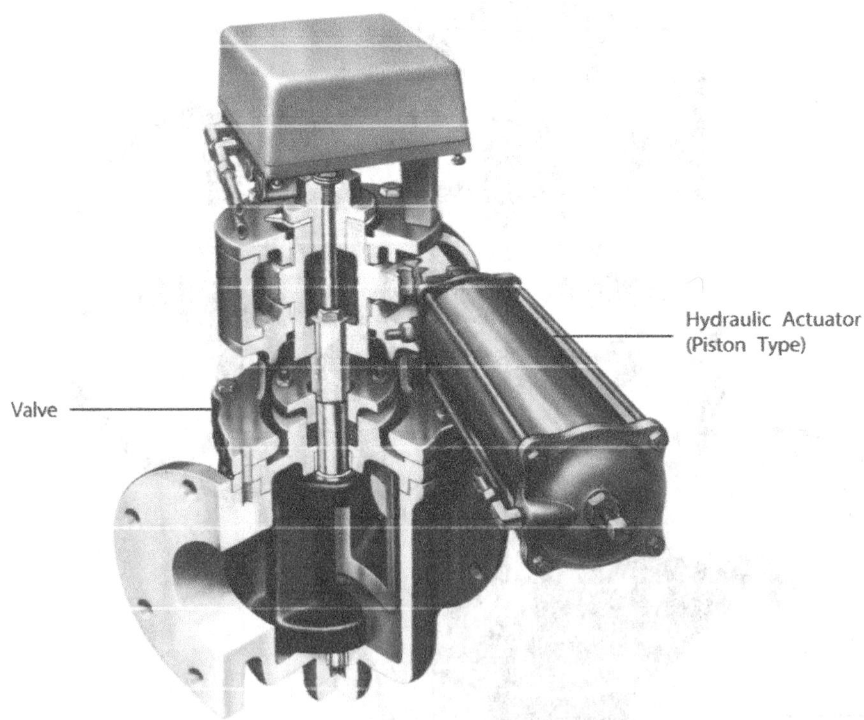

Figure 1-4 Eccentric plug valve with hydraulic actuator

Metal seatings. The material for metal seatings should be carefully chosen because the seatings are prone to damage by corrosion, erosion, abrasion, and deformation. The type of metal chosen for an application should be considered in relation to the types of fluids with which it will come in contact, replacement capabilities, how often it will be operated, and other factors that may cause damage or wear to the sealing mechanism. Different metals offer various sealing abilities and resistance to damage. System designers should choose a valve that has seatings offering the best compromise between sealing ability and wearability based on the environment for which it is designed.

Soft seatings. Soft seatings are sealing mechanisms generally made from various natural or synthetic rubbers or plastics. The soft material readily conforms to the mating surface, creating an effective seal. This type of seating should be designed to prevent the seating material from being moved or deformed by fluid pressure. Presently there are limitations on the use of soft seatings on large valves.

Sealant. Valve passageways can be sealed by a substance (the sealant) injected into the space between the seatings after the valve is closed. The sealant fills any spaces that might be left open by the seatings and thus prevents leakage. Sealant is also used in emergencies to provide a seal when the original seal has failed.

Based on advancements in material and product development, new ways of sealing valves are being implemented and evaluated to enhance the reliability of valves and all their components. Two examples of the newer approaches for coating valves are the fusion-bonded epoxy process as well as the use of the powder coating technology.

This page intentionally blank.

Chapter 2

Types and Selection of Valves

This chapter introduces the major types of valves and the parameters that should be considered when selecting a valve. These parameters include the purpose of the valve, size, operation, and application.

Although certain valve types can be used for many different applications, designers should choose a valve that carefully balances the number of times the valve is operated, maintenance requirements, efficiency, and cost for the specific application. Factors to be considered in the valve selection and specifying process include:

- The requirement that the valve either control or shut off the flow in the manner demanded by service conditions
- The frequency of the valve's operation
- Conformance to appropriate American Water Works Association (AWWA), American Society of Mechanical Engineers (ASME), American National Standards Institute (ANSI), and other product standards
- Ability of the valve to withstand maximum working or test pressures, as set forth in the AWWWA standards
- The need for an unobstructed waterway to maximize flow or accommodate cleaning operations such as "pigging"
- Design and material selection to resist attack by corrosion and/or erosion
- Adaptation requirements for actuator mounting (if any)
- Installation considerations such as weight, accessibility, and dimensional clearances
- Maintenance needs and costs

This chapter will cover only the most common types of valves and their functions. Valves can be grouped based on their method of flow regulation. The general types of flow regulation and the common valves for those types are as follows:

- Flow-control or regulating valves—designed to withstand the demands of usage in the partially open position
  - Globe
  - Piston
  - Diaphragm
  - Butterfly
  - Pinch
- Isolation or shutoff valves—designed to be used primarily in the fully open or fully closed positions
  - Multiturn—gate
  - Quarter turn
    - Plug
    - Ball
    - Cone
    - Duckbill

VALVES USED FOR ON-OFF OPERATION AND FLOW CONTROL

Some valves can be used for both on-off operation and flow control. Most valves are more efficient for either throttling (ball, cone, and globe valves) or on-off (gate valves) applications. The following sections describe the major types of valves and list their best applications. The advantages and disadvantages are somewhat subjective and may not strictly apply in all situations; the information should be used together with engineering data and operational judgment.

Globe Valves

Advantages:

- Suitability for flow regulation
- High sealing capacity
- Resistance to wear

Disadvantages:

- Have a greater ability to allow material to get trapped in the seat
- Produce a higher resistance to flow through the valve
- More prone to reducing the ability to fully close the valve when material is trapped in the seat

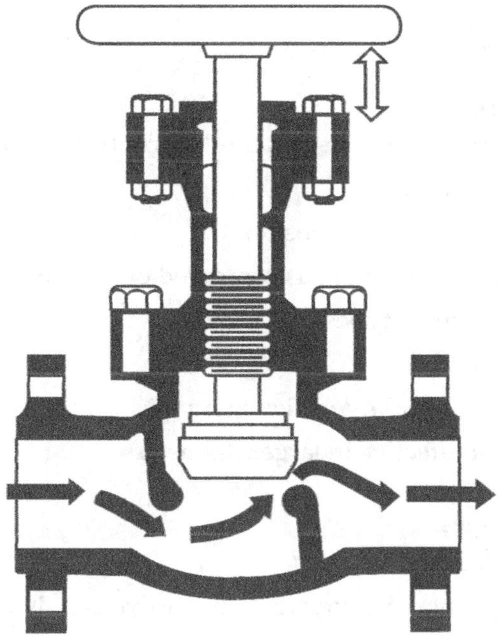

Figure 2-1 Globe valve

Description. Globe valves are regulating valves in which the closure member is moved directly on or off the seat, generally by a rotating threaded stem. The closure member is commonly referred to as a disc. This valve is well suited for flow regulation because of the short distance between the open and closed positions. It also has a high sealing capacity. Because little friction is encountered in opening and closing the seats, the valve is resistant to friction wear. However, globe valves tend to trap particulates in the seat, and the flow resistance is high in these valves. This flow resistance, in turn, usually reduces the flow (see Figure 2-1).

Piston Valves

Advantages:

- Suitability for on-off operation
- Suitability for some flow regulation

Disadvantages:

- High head loss
- Sediment entrapment causing wear

Description. Piston valves use a piston-shaped closure member that intrudes into the seat bore. Sealing is achieved between the sides of the piston and the seat bore. Therefore, flow cannot start until the piston is completely withdrawn. This type of valve is most effective for on-off operation and is somewhat useful for flow regulation. Solids that might be deposited on the seat tend to be wiped away during closing. A primary advantage is that the sealing surface of the piston is protected from flow when the valve is in the fully open position.

Butterfly Valves

Advantages:

- Suitability for both on-off and flow control, 90° operation
- Ease of operation
- Short face-to-face dimension
- Relatively light weight as compared other valve types
- Accept multiple types of actuation
- Lower initial cost
- Can be used for larger-diameter pipes
- More cost-effective in larger diameters

Disadvantages:

- Some head loss
- Partially blocked waterway when valve is fully open
- Incompatibility with pigging (described in this section)
- High-velocity limitations

Description. A common valve in the distribution system, especially in larger sizes, the rubber-seated (sometimes called *resilient-seated*) butterfly valve (see Figure 2-2) has been used in distribution systems since the 1950s. In the butterfly valve, a circular disc or vane that has a diameter less than or equal to that of the pipe is fastened to a shaft running through the valve body and extending outside to an actuator. The disc or vane rotates 90° in the waterway. In the full-open position, the disc is parallel to the flow. The flow, split by the disc, continues on either side of the disc. In the closed position, the disc is perpendicular to the flow and stops the flow. The clearance between the disc and the valve body is sealed by a rubber seat. For additional information on butterfly valves consider AWWA Manual M49, *Butterfly Valves: Torque, Head Loss, and Cavitation Analysis.*

Figure 2-2 Butterfly valve

The rubber seat may be attached to either the disc or the valve body. Butterfly valves are covered by ANSI/AWWA C504, Rubber-Seated Butterfly Valves 3 In. (75 mm) Through 72 In. (1,800 mm); C516, Large-Diameter Rubber-Seated Butterfly Valves, Sizes 78 In. (1,800 mm) and Larger, and AWWA Manual M49, *Butterfly Valves: Torque, Head Loss, and Cavitation Analysis*.

A butterfly valve shuts off pressure as though it were a gasket. The seat holds until the pressure exceeds the seal compression, after which the seal leaks. The valve is set at the factory to hold the rated working pressure (the pressure at which the valve is designed to operate) and should not be expected to seal at higher pressures. The reduction in flow of the butterfly valve is a function of the angle of the disc opening in the pipe. The valve is generally effective for flow control in the range of 20° to 70° open, depending on the difference in pressure across the valve and other characteristics of the valve. Butterfly valves may be used for throttling in some situations; however, they are not intended for long-term throttling. The designer should consult the manufacturer before specifying a butterfly valve for throttling applications.

Advantages of butterfly valves are that they provide shutoff capabilities, some throttling capabilities, and ease of operation, particularly in sizes 12 in. and larger. The actuators required by larger valves increase the mechanical advantage to allow for ease of operation. Additionally, butterfly valves that are 12 in. and larger generally have a lower initial cost than other valve types.

However, butterfly valves do not provide a clear waterway. Restriction of flow in a fully open butterfly valve is greater than in some other types of valves because the disc remains in the waterway. As the valve size increases, the proportion of the open flow area to the cross-sectional area of the disc increases. If a large number of butterfly valves are used, pressure loss in a distribution system can be significant. When flow resistance is critical, other types of valves may be required.

Another disadvantage is that butterfly valves prevent the use of pigging, which involves using line pressure to force a bullet-shaped plug through a water line to locate and clean flow restrictions.

Plug Valves

Advantage:

- Suitability for both throttling and flow regulation

Disadvantage:

- Cause some flow resistance

Description. Plug valves (Figure 2-3) are primarily used in systems where slurries, grit, or solids will flow through the valve. The eccentric plug valve is a quarter-turn valve allowing a cost-effective low-torque actuation for shutoff and throttling applications. These valves can be adapted to operate with various types of manual or automated actuation. Plug valves are available in flanged, mechanical joints, or threaded-end configurations in standard or 100 percent ported design per ANSI/AWWA C517.

Flow restriction in a plug valve depends on the size and shape of the plug opening. For example, some plug valves have tapered openings where the opening is smaller at the bottom than at the top. Tapered plugs cause more flow restriction than a rectangular design of the same size because there is more plug area blocking the flow. Flow resistance can range from that which is slightly more than a gate valve to more than that of a butterfly valve. Advantages of the plug valve include that it can be installed in any position without special consideration and, when utilizing a rectangular port design, change in flow is consistent throughout travel of the plug.

Figure 2-3 Eccentric nonlubricated plug valve

Larger plug valves require an actuator for 90° rotation. Smaller plug valves may be operated with either a 90° rotating lever handle or actuator.

Conically tapered metal (often *brass*) plug valves are often used as simple shutoff valves in household *natural gas* lines.

Cone Valves

Advantages:

- Full waterway opening with low flow resistance
- Good flow-control characteristics
- Suitability for throttling or flow modulation

Disadvantages:

- Large physical size and weight require facility to be preplanned to accommodate the valves
- More costly when compared to other types of valves, such as ball, gate, or butterfly valves
- Maintenance of this valve is more difficult and costly than maintenance of some of the other valves

Description. The cone valve (Figure 2-4) is a quarter-turn (90°) rotary valve for which the flow-control device is a cone seating in a conically shaped body. The seating surfaces are commonly metal to metal. Cone valves have a tolerance for high velocities. Cone valves are typically used to control flow and in pump discharge where high head and high velocities occur; this can require single or multiple configurations. These valves can work well in areas where controlling the variations of flow or where higher velocities within the system are experienced. These valves are often used in dam structures, hydroelectric facilities, and water treatment plants and systems, as well as in spillways.

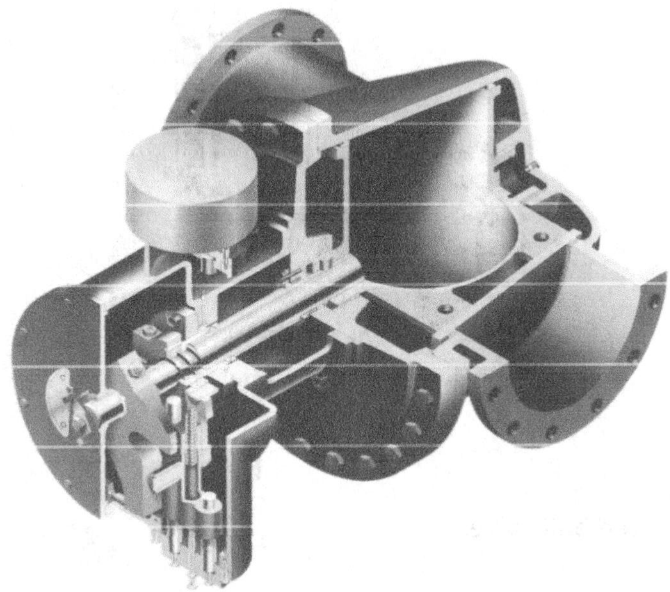

Figure 2-4 Cone valve

Ball Valves

Advantages:

- Low flow resistance
- Suitability for throttling where pressure reduction is important
- High-pressure capability
- 100 percent flow area equal to pipe size
- Energy savings caused by less friction loss
- Adaptability available with metal-to-metal or resilient seating
- Multiple types of actuators can be used

Disadvantages:

- In larger-diameter ball valves, the physical size and weight require facility to be preplanned to accommodate the valves
- Higher initial cost
- Do not offer the best flow control option in every situation
- More susceptible to water hammer

Description. The flow-control mechanism in the ball valve is a sphere that has a waterway opening. The sphere is connected to shafts perpendicular to the waterway, and there is 90° of rotation from the closed position to open position. The seating surfaces are either resilient on a noncorrosive metal or noncorrosive metal to metal. Valves are capable of bidirectional pressure ratings because when ball valves are furnished with port openings of the same size as the pipe opening, they have a very low head loss (see Figure 2-5). Ball valves are typically used in pump discharge, high head, high velocity applications

Figure 2-5 Resilient-seated ball valve

where typical flow control and antisurge control are critical. These valves are adaptable with manual actuation, electric motor actuation, or hydraulic cylinder actuation, depending on the specific application.

Small rubber-seated ball valves are used in consumer service connections as corporation stops, curb stops, and meter isolation valves. Larger ball valves are generally installed in applications where the pressure or velocity exceeds the capability of a butterfly valve or gate valve. Ball valves are used extensively for control or throttling service where a moderate reduction in pressure or flow control to reduce the potential for water hammer is required. In such service, the actuator is constructed to provide a mechanical advantage for minimal torque requirements so that the valve operates smoothly and easily. Ball valves are covered by ANSI/AWWA C507, Ball Valves, 6 In. Through 60 In. (150 mm Through 1,500 mm), and by ANSI/AWWA C800, Underground Service Line Valves and Fittings.

Pinch Valves

Advantages:

- Ease of operation
- Resistance to corrosion

Disadvantages:

- Effective for flow control only when they are at least 50 percent closed
- Typically more expensive than many other types of valves
- Used generally for small pipe size applications
- Do not allow the effective operation of the cleaning or pigging process

Description. Pinch valves are simple in design. They shut off flow by pinching and sealing off a synthetic or natural rubber tube. They are similar to the pinch cock used in laboratories. The operating mechanism is separated from the fluid, so these valves are useful where corrosion and contaminated fluids are a problem, which should not be a consideration in water distribution systems (see Figure 2-6).

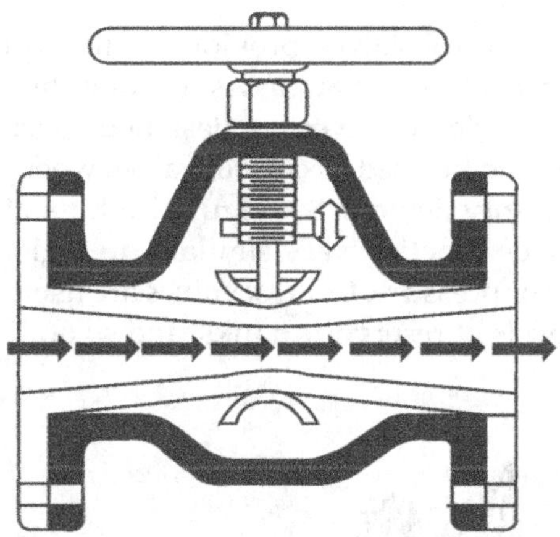

Figure 2-6 Pinch valve

Diaphragm Valves

Advantages:

- Ease of operation
- Resistance to corrosion
- Resistance to erosion

Disadvantages:

- Only effective for flow control after 50 percent closed
- More expensive than most other valves
- High head loss
- Generally only available in small sizes
- Not effective for the cleaning or pigging process

Description. Diaphragm valves (see Figure 2-7) contain a flexible diaphragm attached to a compressor. The diaphragm is lowered by a valve stem onto a weir, which seals and shuts off the flow, similar to how a pinch valve operates. Diaphragm valves are used to handle corrosive, erosive, or contaminated fluids, which should not be a consideration in water distribution systems. Their primary application and use are in industrial environments as well as in some water and/or wastewater treatment facilities.

Control Valves

Advantages:

- Accurate proportioning

Disadvantages:

- Possible cavitation resulting in excessive wear if wrong type of valve used
- High head loss
- Limited application and/or use
- Require a lot of maintenance to be effective

Description. Control valves provide accurate proportioning control by automatically varying the rate of flow that passes through the valve according to signals received from sensing devices. Some valves are designed specifically as control units. Most types of valves, however, can be used as control valves when combined with actuators, positioners, and other accessory devices. See Figure 2-8 for a photograph of a typical flow-control valve. Control valves function very similarly to a globe valve and are typically used to control either flow or pressure. Control valves are used typically to manage flow, pressure, and the various levels of reservoirs, tanks, and so on.

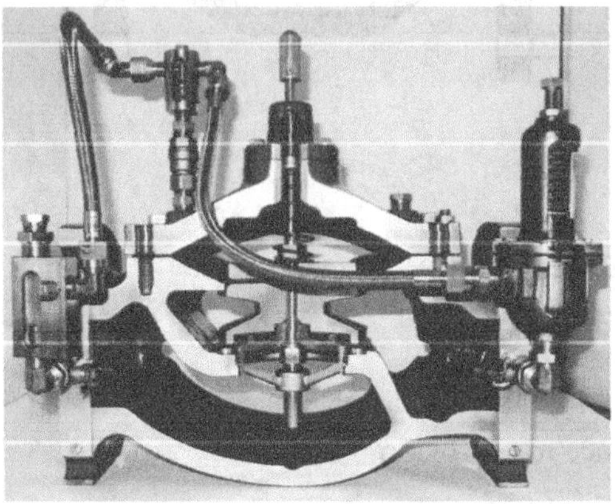

Figure 2-7 Diaphragm valve

Figure 2-8 Flow-control valve

VALVES USED PRIMARILY FOR ON-OFF OPERATION

The valves discussed in this section are used primarily for on-off control of water, rather than for flow control.

Gate Valves

Advantages:

- Clear waterway for low flow resistance
- Bidirectional drip tight seal
- Lower pumping cost
- Low to no restriction/friction (i.e., low head loss)

Disadvantages:

- Some designs are not suitable for flow regulation
- Metal-seated valves may require more frequent exercising than resilient-seated gate valves
- Are not designed for throttling applications
- Can be more expensive in larger sizes
- Size consideration/additional expenses (for instance, larger-size valves are more cumbersome and harder to move around, and bigger equipment must be used to get them in place)

Description. Gate valves are the most commonly used type of valve for isolating portions of the distribution system. Gate valves block the passage of water with a disc sliding perpendicular to the flow. Primarily used for open or closed service, gate valves are typically not recommended for flow regulation or for throttling. Gate valves are not well suited for precise flow control because flow reduction is not proportional to travel (the extent to which the valve is open); very little flow resistance occurs until the valve is about 75 percent closed. Large gate valves should never be utilized in a severe throttling application given cavitation forces created by compressed flow under the gate. In the full-open position, the gate has a clear waterway that keeps flow restriction to a minimum. See Figure 2-9 and Figure 2-10 for illustrations of gate valve operation. The most common type of gate valve is the resilient-seated gate valve. Older style designs, including double-disc and solid-wedge gate valves, are rarely used today and have limited availability.

Double-disc gate valves. Covered by ANSI/AWWA C500, Metal-Seated Gate Valves for Water Supply Service, double-disc valves have been in general use since the early 1900s. The design consists of two relatively loose fitting discs that, when closed, are pressed against metal seats by a wedging mechanism. The wedging force is released when the valve is opened. Traditional double-disc valves are not recommended for throttling or flow-control applications.

With double-disc valves, water pressure pushes past the upstream disc and allows the area between the two discs to become pressurized when the discs are in the closed position. This pressure then forces the downstream disc against the seat. Generally, the higher the pressure, the tighter the seal. Larger valves are usually equipped with a bypass valve for pressure equalization to ease main valve operation. The addition of gearing is common on valves 16 in. and larger to reduce the effort required to operate the valves. If valves 16 in. and larger are installed horizontally, rollers, tracks, and scrapers should be

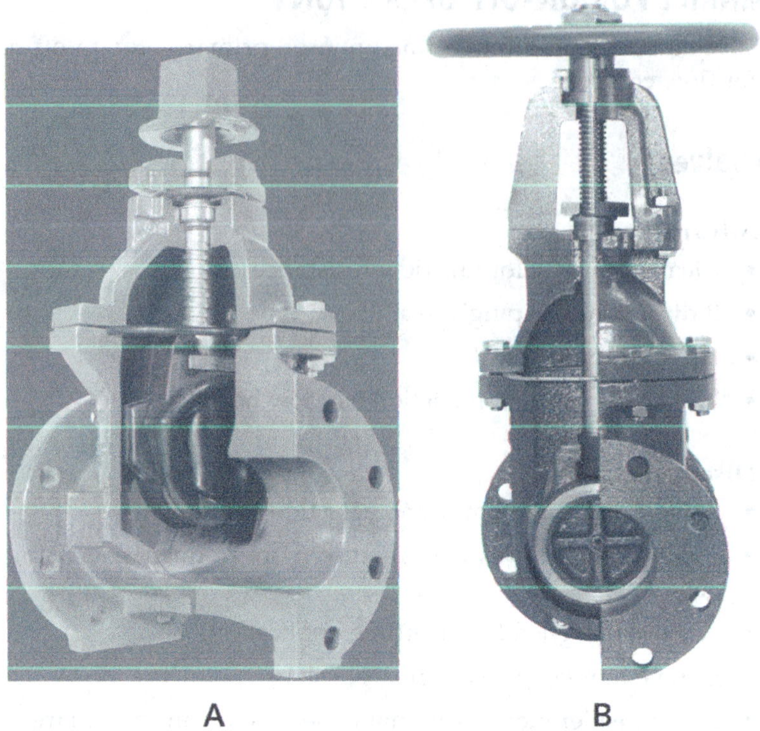

Figure 2-9 Gate valves: (A) Solid wedge and (B) resilient wedge

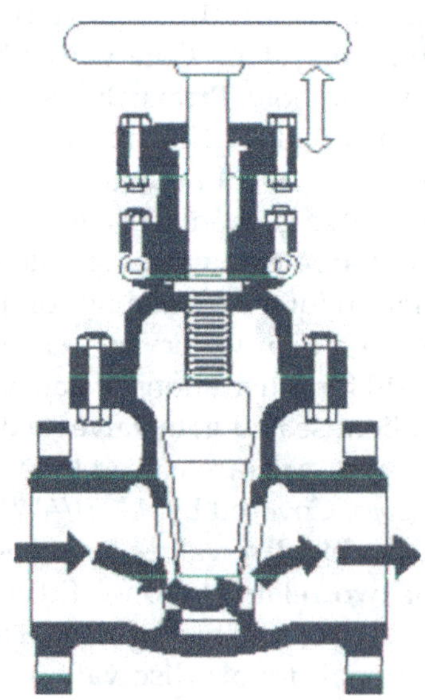

Figure 2-10 Gate valve

used to make operation easier. Fluids with any debris (grit or mineral deposits) clog these tracks over time. Gate valves that have revolving discs do not require rollers.

The advantage of double-disc valves is that they are bidirectional, providing flow shutoff in either direction. A principal disadvantage of double-disc valves is the large frictional force that must be overcome to move the downstream disc off the seat. Two people or an actuator may be required to operate the larger valves (more than 12 in. in size). Also, when using an actuator, the operator must be concerned with excessive torque that may damage the valve system. In some waters, corrosion may damage the valve and make it more difficult to operate.

Solid-wedge gate valves. Also covered by ANSI/AWWA C500, the solid-wedge design involves a closure member that is a single wedge guided into fitted, tapered body seats. The wedge seats do not contact the body seats until final closure. High pressure may cause the upstream seat to allow water into the valve body but, as with the double-disc valve, the downstream seat will seal even tighter with additional pressure.

Although this type of valve is best suited for isolation or full-open use, the wedge gate may be somewhat more acceptable than the double-disc valve in a throttled position because of the close guiding between the wedge and body. This type of valve also provides a clear waterway that minimizes flow restriction.

Resilient-seated gate valves. Resilient-seated gate valves are used to cover both gray-iron and ductile-iron bodied valves, which are now commonly referred to as the full-wall valve standard. Resilient-wedge valves are the most common type of resilient-seated valves being produced today.

In an effort to reduce the weight of larger valves being introduced into water systems, a newer standard—ANSI/AWWA C515, Reduced-Wall, Resilient-Seated Gate Valves for Water Supply Service—was developed to cover ANSI/AWWA Standard C509, Resilient-Seated Gate Valves for Water Supply Service. From the new standard "reduced-wall" gate valves were produced in ductile iron only. By separating "full-wall" and "reduced-wall" valves into different standards, purchasers are now able to clearly identify their preferences. The resilient-seated gate valve has a closure member consisting of a gate that may be encapsulated with a resilient material (rubber, urethane rubber, or others). ANSI/AWWA C509 also allows designs for which the resilient seat is attached to the body. The first illustraton in chapter 3, Figure 3-1, shows a sample shop drawing of a resilient-seated gate valve with an outside stem and yoke.

These valves are bidirectional and provide a clear waterway when open. Because of their smoother waterway, the head loss is usually less than the head loss for metal-seated gate valves. These valves should not be used for throttling because throttling may damage the valve.

SPECIAL-PURPOSE VALVES

For specific system applications, valves are available to make operation simpler, more efficient, or automatic. Special-purpose valves are designed to prevent backflow, automatically control pressure, release air, or allow tapping of water lines without isolating flow.

Check Valves

The check valve is a single-direction valve that allows flow in one direction (forward) and stops reverse flow (see Figure 2-11). It opens under the influence of forward pressure and flow, and it closes automatically when flow ceases or reverses, making operation semiautomatic. In some check valve designs, an external lever with a weight or lever spring helps close the valve prior to flow reversal. Check valves are used in pump discharge and flow applications where it is of importance to stop the reverse flow of a liquid prior to pump and/or in water column reversal in order to reduce the chance of water hammer. Typical applications include treatment plants, pump stations, and lift stations.

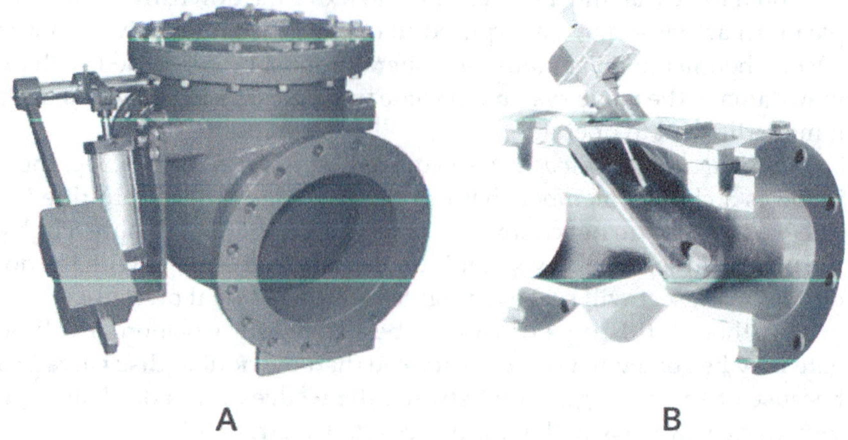

Figure 2-11 (A) Swing-type check valves with lever and weight and (B) air cushion device/standard swing-check per C508

Many other types of check valves, such as swing-check valves, ball check valves, and others, are available; but these are considered specialized valves and will not be discussed here. Horizontal swing-check valves are covered by ANSI/AWWA C508, Swing-Check Valves for Waterworks Service, 2-In. (50-mm) Through 24-In. (600-mm) NPS.

Air Valves

Air valves are used in transmission and distribution pipelines to increase operating efficiency and provide pipeline protection. They are generally divided into three types of valves: air-release, air-vacuum, and combination air valves per ANSI/AWWA C512, Air-Release, Air/Vacuum, and Combination Air Valves for Waterworks Service, and AWWA Manual M51, *Air-Release, Air/Vacuum, and Combination Air Valves.*

Air-release valves. An air-release valve (see Figure 2-12) is generally a self-actuated valve that automatically vents small pockets of air that accumulate at the high point in a water distribution system when the system is operating under pressure.

After an air-release valve is installed at the high point in a system, it fills with water and closes. During system operation, small amounts of air enter the valve from the system. As air displaces water within the valve, the water level drops. When the water level falls to the point where the float associated with the valve is no longer buoyant, the float drops. This action opens the valve orifice and allows the air that has accumulated in the upper portion of the valve body to be released into the atmosphere. As this air is released, the water level within the valve rises, lifting the float and closing the valve orifice. This cycle repeats itself whenever air accumulates in the valve.

The ability of the valve to open and release accumulated air under pressure is achieved through the use of a leverage mechanism. As the water level and float drop, the weight of the float acting on the mechanism produces a greater force to open the valve than the system pressure exerts to hold the valve closed. Generally, the higher the system pressure, the smaller the orifice diameter needed in the air-release valve.

Air-vacuum valves. Air-vacuum valves (see Figure 2-13) are often referred to as *large-orifice air valves*. The valve is designed to allow large volumes of air entrapped in the pipeline to escape during filling. Once the float is closed, the valve will not operate again until a negative pressure occurs within the pipeline caused by draining or a column separation. When this occurs, the float will drop open to allow large volumes of air to enter the pipeline, preventing the loss of pressure, which could cause the pipeline to collapse and/or which could cause damaging water column separation.

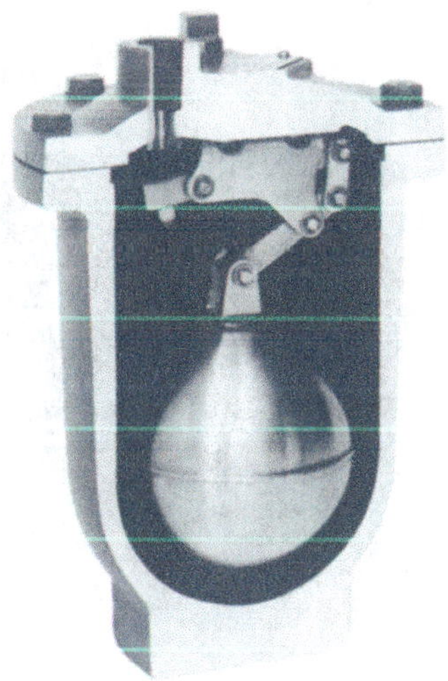

Figure 2-12 Air-release valve

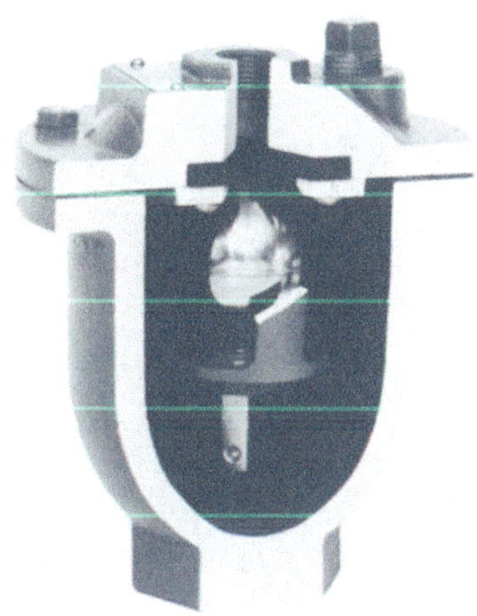

Figure 2-13 Air/vacuum valve

Combination air valves. Combination air valves (see Figure 2-14) incorporate the use of both a small-orifice air-release valve and a large-orifice air-vacuum valve either in a single-housing or dual-body design. The combination valve provides the function of both an air-release and an air-vacuum valve. When in doubt of which valve type to use, the combination valve will ensure the proper protection and efficiency of the pipeline.

Inflow Preventer

Water distribution systems are equipped with air valves at multiple locations to release and admit air into pipelines and reservoirs to allow filling and draining, control surges, and maintain flow efficiency. If the valve or reservoir outlets become flooded or the target of malicious tampering, the air entry point may be compromised and allow the entry of flood water or the intentional introduction of a contaminate into the water system.

Figure 2-14 Combination air valve

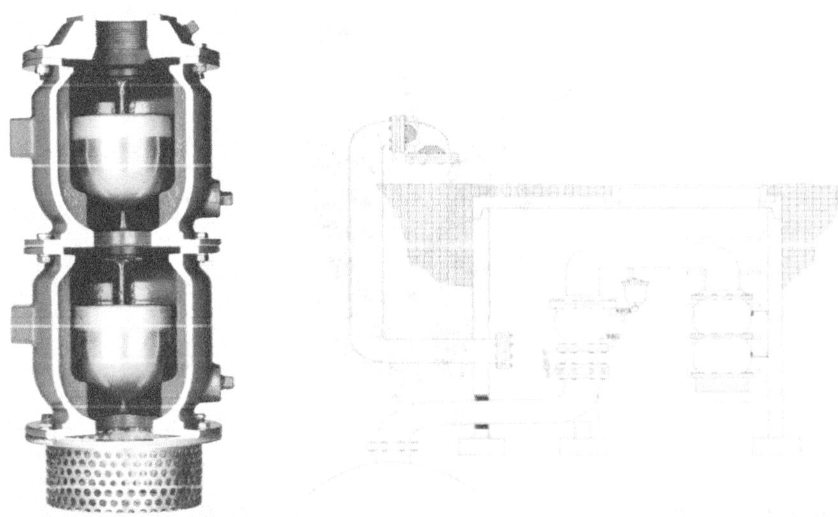

Figure 2-15 Flood-safe inflow preventer

The inflow preventer (see Figure 2-15) is directly connected to air-valve outlets in valve vaults and reservoir vent pipes. This device is normally open to allow unrestricted airflow in and out of pipelines and reservoirs during normal operating conditions. When a flood occurs in a valve vault or reservoir vent, a float-operated closure member automatically closes to prevent entry of the contaminated fluids. A redundant float-operated closure member automatically closes if the first one fails.

Altitude Valves

Altitude valves are mostly globe-type body valves normally installed in storage tank inlet-outlet lines; they remain open until the tank is filled. The valves are then closed during normal flow conditions. They open again when the pressure in the distribution system becomes less than the static head of the height of water in the tank. These valves are usually equipped with a hydraulic pilot system that senses pressure and operates the valve.

Pressure-Relief Valves

Pressure-relief valves are used to protect against excess pressure in water lines. They lower the pressure by releasing water when the designated pressure safety limit is exceeded. They close again when the line pressure falls below the safety limit (see Figure 2-16). These types of valves are commonly used to relieve high pressure conditions.

Pressure-Reducing Valves

Pressure-reducing valves are used to provide water to a pressure district (or zone) of lower elevation from a district of higher elevation. They are often globe valves equipped with hydraulic pilot systems similar to those of altitude valves. The valves do not close under normal flow conditions, unless doing so is required to shut off flow. Typically, a large-volume valve and small-volume valve are used in parallel because of the possibility of a wide variation of pressures to be reduced. A photograph of a pressure-reducing valve is shown in Figure 2-17.

Tapping Valves

Valves used for tapping service are usually resilient-seated gate valves, which are covered by ANSI/AWWA C509 and C515. Tapping valves are designed to be used in conjunction with a tapping sleeve or tapping saddle and a standard tapping machine. Operators can install the sleeve and valve and make the tap under pressure in a takeoff line from an existing main without shutting down or otherwise interrupting the flow of water in the main. To ensure the proper installation of the tapping valve and tapping sleeve, the assembly should be tested with water with the gate closed. This test will determine if the saddle is watertight prior to drilling the water main. The test should be conducted prior to the installation of the tapping machine and prior to making the tap. Under no circumstances should the assembly be tested with air. If air is used, the operator may face serious injury or even death.

One end of the tapping valve consists of a modified flange with a raised pilot. This flange mates with the recess or counter-bore in the corresponding flange of the tapping sleeve or saddle. This feature enables proper alignment of the valve on the sleeve or saddle and ensures that the shell cutter of the tapping machine meets no interference as it passes through the valve and saddle. The mating dimensions are standardized by the Manufacturing Standardization Society (MSS) in Standard Practice 60, Connecting Flange Joint Between Tapping Sleeves and Tapping Valves.

The other end of a tapping valve attaches to the tapping machine. This end can be furnished with a flange, a standardized mechanical joint, or a proprietary push-on joint. The interface dimensions for tapping valves and tapping machines for nominal sizes up to 48 in. are described in MSS Standard Practice 60. When selecting a tapping valve, operators should determine whether the valve can be closed without the gate striking the retracted pilot drill; they should also make sure the valve is compatible with the tapping machine. The distance from the tapping machine to valve interface to the valve gate is not standardized among valve manufacturers. Tapping machine travel should be determined by the tapping machine using the size of the pipe being tapped with the machine being used.

Other pertinent MSS standard practices include SP-111, Gray-Iron and Ductile-Iron Tapping Sleeves, and SP-113, Connecting Joint Between Tapping Machines and Tapping Valves.

Figure 2-16 Pressure-relief valve

Figure 2-17 Pressure-reducing valve

Insertion Valves

An insertion valve (Figure 2-18) is a specially designed gate valve that, when used with suitable valve-insertion equipment, can be installed in an existing pipe system under full operating pressure without interrupting service. Insertion valves are placed on a previously unvalved line to control flow. Some valve sizes require a companion sleeve. These valves are produced in sizes 4 in. through 54 in. for installation in cast-iron or ductile-iron mains. They may also be used with certain sizes of steel pipe, concrete pressure pipe, and asbestos-cement pipe, depending on the wall thickness and the outside diameters of the pipe. Because of the specialized equipment required to use an insertion valve, many utilities hire a contractor to perform the work.

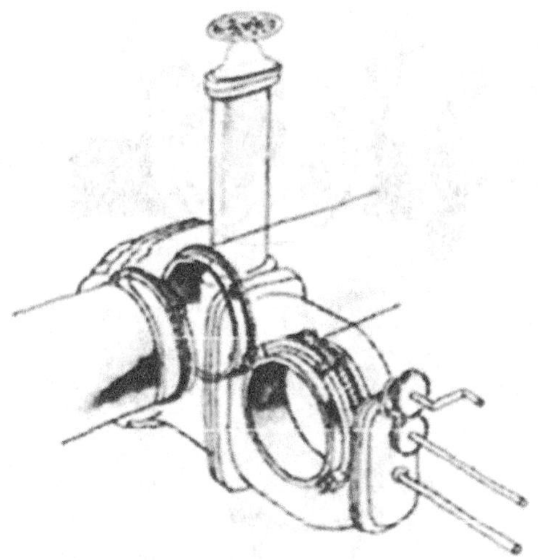

Figure 2-18 Insertion valve

ACTUATORS

A manual actuator is any lever, gear, or wheel used to facilitate movement of a valve's control element as shown in Figure 2-19. An automatic power actuator uses an external power source that provides the torque (turning or twisting force) necessary to operate a valve automatically or remotely. Automatic power actuators are often used on valves that are frequently operated; they are a necessity on pipeline valves in remote areas. Certain power actuators can quickly shut down a pipeline in case of an emergency. Refer to chapter 1, Figure 1-4, for an illustration that shows a plug valve with an actuator attached.

Hydraulic and Pneumatic Actuators

Hydraulic actuators (see Figure 2-20) and pneumatic actuators are often simple devices, with a minimum of mechanical parts. These actuators are used on quarter-turn valves and other rotary devices. Air or fluid pressure provides the required thrust or torque, acting on a piston linked to the valve stem to open or close the valve. Both types of actuators can be supplied with a fail-safe feature that turns the valve on or off as required if there is a sudden drop in operating pressure or a power failure. Because air is compressible, it should not be used on high-torque applications or where smooth operation of the valve is essential. If these actuators are installed underground, a vault should be installed that will allow future maintenance.

Electric Actuators

Electric actuators have motor drives that provide the thrust or torque to operate a valve. Electric actuators (see Figure 2-21) are an alternative to air-type or fluid-type actuators. They are most frequently used on multiturn valves and on quarter-turn valves. Finally, the valve actuator may require maintenance or repair periodically. It should be placed in a vault or an easily accessible location.

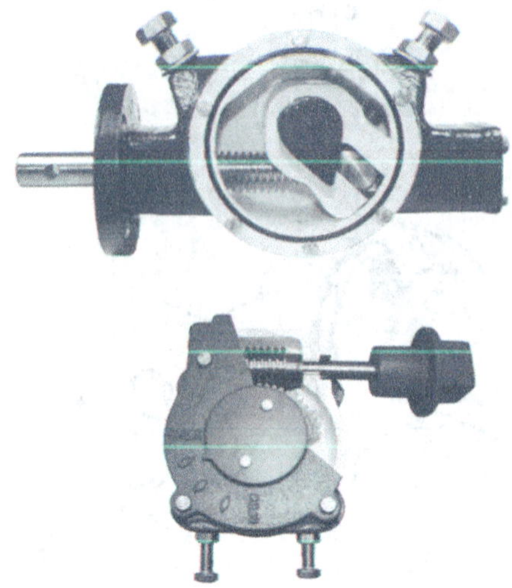

Figure 2-19 Manual actuators/traveling nut/worm gear

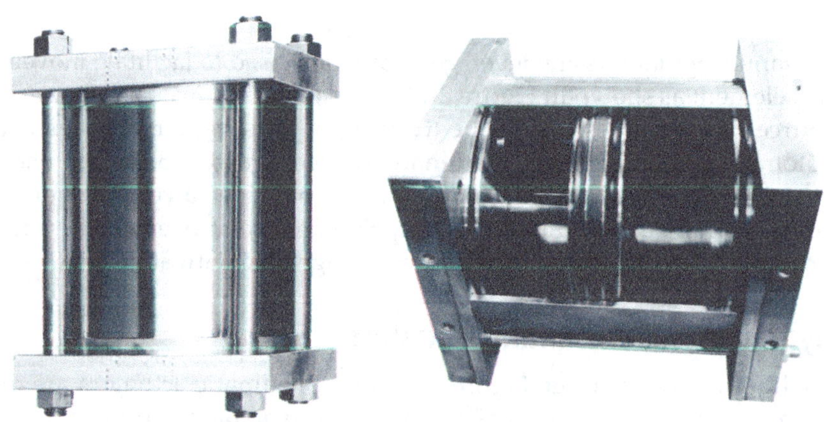

Figure 2-20 Hydraulic cylinder actuator

Figure 2-21 Electric actuator

VALVE-END CONNECTIONS

Valve ends of the mechanical joint, proprietary push-on joint, or flanged joint types are commonly available and can be found in combination. Other valve ends such as threaded, caulk hub, shoulder joint, and straight-spigot types are typically less available. Valve ends (joints) must be compatible with the adjacent piping component and should be installed properly to achieve a leak-free joint.

In addition to flanged joint ends, special restrained push-on and mechanical joint designs are available with one or more of their joint components modified to provide mechanical locking between the valve end and the pipeline.

Detailed design information, pressure ratings, and assembly procedures can be obtained from the valve manufacturer. Some joints are proprietary, and the manufacturer has exclusive rights to their designs. Other information on joints can be obtained from the appropriate AWWA standard. Some typical valve-end connections are shown in Figure 2-22.

Valve body material is typically stainless steel, ductile iron, or cast iron. Wall thicknesses vary and should be analyzed. Valve body coatings should also be analyzed for specific use and durability. Bolts and nuts are typically zinc-coated mild steel. Many manufacturers will use stainless steel or other types of material for bolts and nuts on valves if specified by the user. Appropriate measures should be taken to protect the valves, bolts, and nuts against corrosion. A typical approach to this is the use of polywrap secured completely around these components.

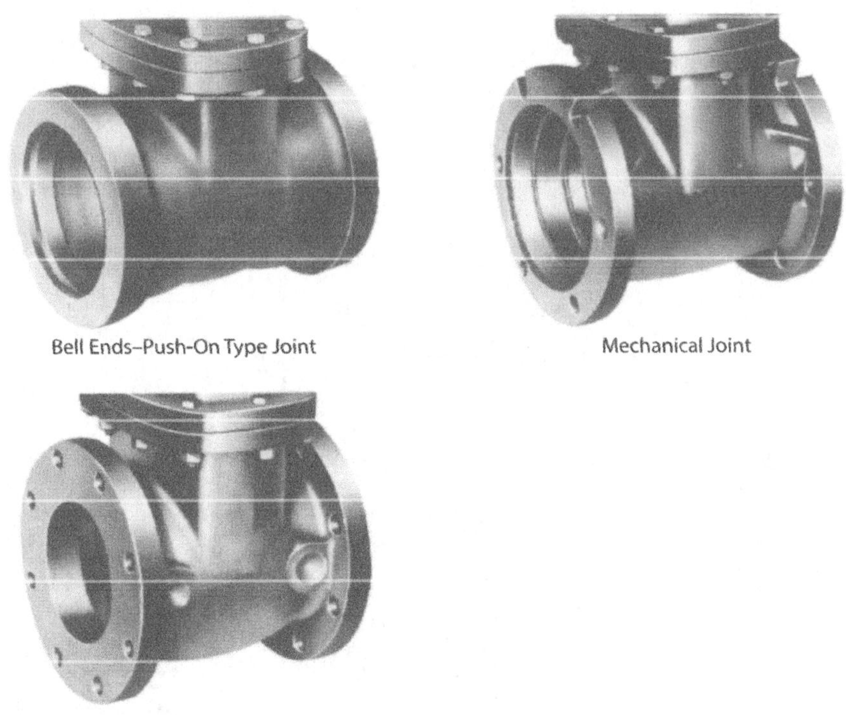

Bell Ends–Push-On Type Joint

Mechanical Joint

Flanged Joint

Figure 2-22 Valve-end connections

This page intentionally blank.

Chapter 3

Valve Tests, Unloading, Inspection, and Storage

Most valve components are manufactured to standards set by the American Water Works Association (AWWA) or other organizations such as the American Society of Mechanical Engineers (ASME) and the American National Standards Institute (ANSI). These standards typically include test requirements or testing procedures to ensure that the valves are manufactured to specifications, and the manufacturer must meet all such requirements and procedures. Valve inspection at the manufacturing facility ensures that valves meet those standards and that the valves are in working order when they are shipped. However, valves should also be tested operationally when they arrive at the utility and before they are installed.

MANUFACTURING PLANT VALVE TESTS

When valves are manufactured, the work is available for inspection and approval by the purchaser's duly authorized inspector. Notice of this requirement should be in the purchaser's valve specifications. In-plant inspections provide additional assurance to the purchaser that the valves meet the requirements of the specification. For in-plant inspections, the manufacturer provides full facilities for inspection and observation of tests. The purchaser's inspector should have access to all places of manufacture where materials are being produced or fabricated or where tests are conducted. An affidavit of compliance with the appropriate product standard may be required from the manufacturer even if the purchaser has an inspector at the plant.

After manufacture, each valve should be subjected to operation and hydrostatic tests (pressure tests of the valve to ensure that there is no leakage) at the manufacturer's plant as specified in the appropriate product standards. Testing assures the buyer that the valve will perform properly. Tests conducted by the manufacturer vary based on valve type and product standard. Some of those tests are described in the following paragraphs.

Operation Test

Each valve should be operated one or more times in the position for which it was designed to ensure correct functioning of all parts. If the valve does not meet standards, it must be rejected. All defects in assembly or materials must be corrected and the test repeated until satisfactory performance is demonstrated.

Valve Body Test

With the valve in the open position, a hydrostatic test pressure equal to twice the rated working pressure (or that specified by the product standard) of the valve should be applied to the body. No leakage is permitted through the metal, joints, or stem seals.

Valve Seat Test

A test should be made from each direction at the rated working pressure (the pressure at which the valve is designed to operate) for the time specified by the standard to prove the sealing ability of each valve from both directions of flow. No leakage is permitted through the metal, through the seats, or past the seat, unless otherwise indicated in the specific standard. The seats of check valves are tested in the stopped-flow direction.

Valve Torque Test

A sample of each valve should be torqued in both the open and closed positions to ensure that there is no stem damage or seating damage and a resulting failure to seal. The specific product standard for each valve type specifies the amount of torque that should be applied. Deviations in the methods used in conducting these tests are specified in the appropriate AWWA standard.

MANUFACTURER'S DOCUMENTS

The purchaser may specify that the manufacturer provide documents to help the buyer properly use the valves. The documents may include an affidavit of compliance with the appropriate product standards, shop drawings, and installation, operations, and maintenance manuals to show dimensions, construction, and materials used. This information is vital to ensure proper installation and operation of the valves.

Shop Drawings

The buyer may request that the manufacturer provide assembly drawings (shop drawings) for all valves delivered. The drawings should include clearance requirements, dimensions, and construction type. They should be sufficiently detailed to show that the valve meets the requirements to comply with standards. Figure 3-1 shows a sample shop drawing. The drawings are especially important when the number of turns, input torque, actuator data, and protective coating are not specified in the standard. Knowledge of the exact number of turns and direction to open or close a valve is essential to the correct operation of the valve (isolation or modulation) as well as to prevent actuator failure.

Installation, Operations, and Maintenance Manuals

Installation, operations, and maintenance manuals from the manufacturer should be provided for each significant order and for all valves. The manuals should include specific instructions for assembly, installation, and operation. The manuals should include suggested maintenance procedures for the operation of valves in service. Proper maintenance can then be scheduled to maximize the service life of the valve and ensure proper operation. Many manuals also include troubleshooting tips and repair instructions. Often actuators can be adjusted in the field to improve closure. Actuators can be damaged by over-torquing. This should be a consideration when selecting actuators and valves. Figure 3-2 shows some typical information from an operations and maintenance manual.

Packaging and Marking

The manufacturer is required to package valves for shipment to prevent damage in moving the materials to the utility and to mark valves with information useful to the user. As detailed in the appropriate standards, such markings may include the name of the manufacturer, the rated working pressure, the year of manufacture, and the size of the valve. Serial numbers on large valves (12 in. and larger) should be recorded in utility records to allow parts to be ordered in the future.

UNLOADING AT THE UTILITY

All valves should be unloaded carefully and lowered gently from the truck to the ground, not dropped. The expanded use of pallets, shrink wrap, side-load trucks, and forklifts has simplified receipt and unloading of large quantities of valves. However, the handling of single valves or uncrated valves is still prevalent and the proper use of rigging (slings, chains, or strapping) is critical. In the case of larger valves, forklifts under the skids or slings around the body of the valve should be used for unloading. Only hoists and slings with adequate load capacity to handle the weight of the valve should be used. Hoists should not be hooked into or chains fastened around stems, bypasses, yokes, gearing, motors, cylinders, or handwheels.

To prevent damaging or chipping special coatings, such as epoxy, valves should be protected from lifting devices with padding. If a special coating is damaged, it should be repaired with the original coating material or a material specified by the coating manufacturer.

Some manufacturers make special lifting devices that minimize valve damage during movement or installation of valves.

Resilient Wedge Flanged OS & Y Gate Valves

GENERAL SPECIFICATIONS

- Resilient-Seated Gate Valves to AWWA C509 (OS&Y)
- Flanged Ends to ANSI B16.1-1975 class 125
- Working Pressure 200 psi
- Test Pressure 400 psi

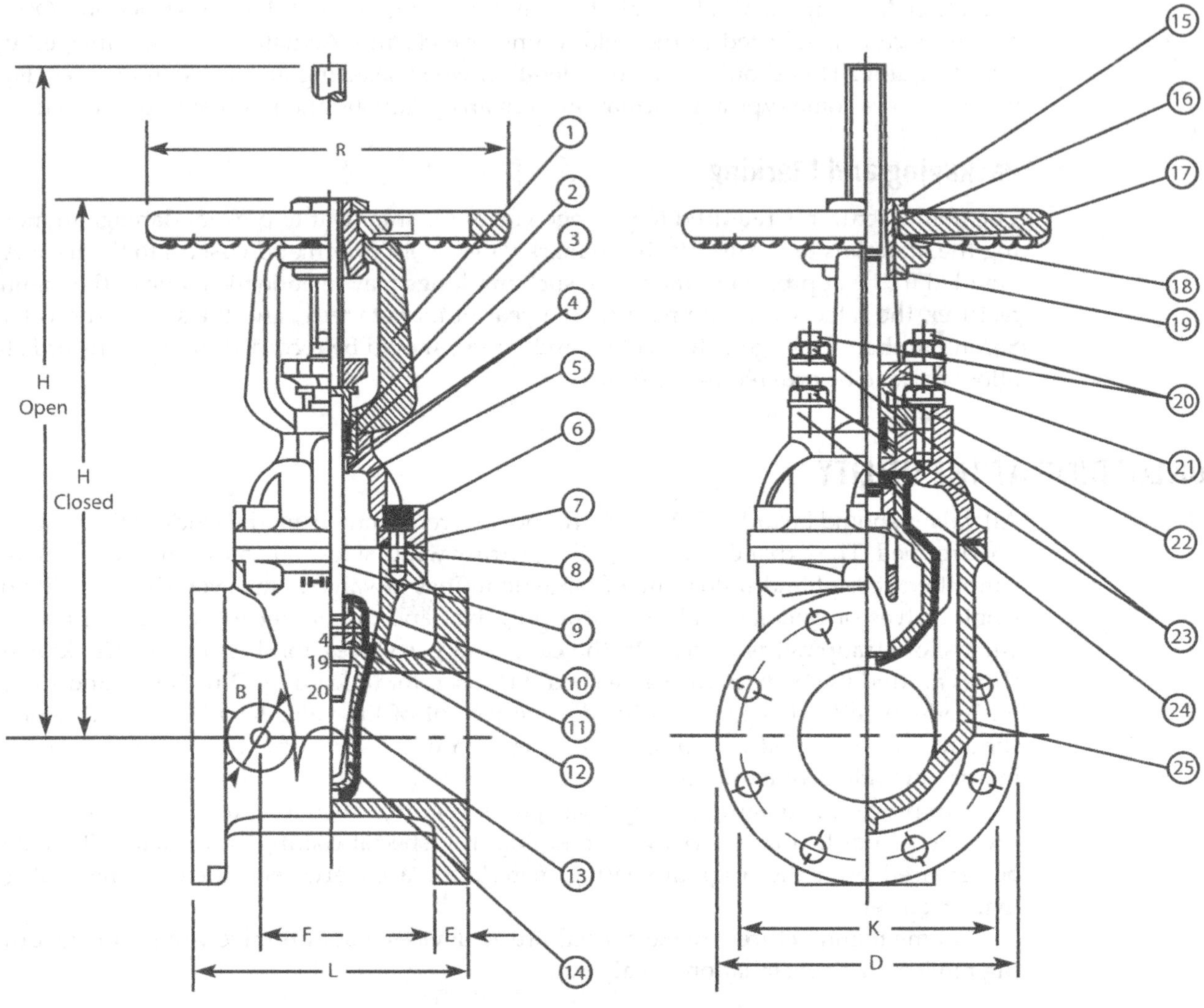

Ref Nos.	H Size	H Open	Closed	R	L	E	D	K	Holes	F	B	Turns to open	Weight lb
25-065-46	2.5″	15.66″	13.06″	9.00″	7.50″	0.71″	7.00″	5.50″	4	3.94″	2.0″	17	50
25-080-46	3″	16.88″	13.78″	9.00″	8.00″	0.79″	7.50″	6.00″	4	4.33″	2.0″	20	55
25-100-46	4″	20.63″	16.53″	11.00″	9.00″	0.98″	9.00″	7.50″	8	4.96″	2.0″	21	85
25-150-46	6″	28.98″	22.88″	13.00″	10.50″	1.06″	11.00″	9.50″	8	6.22″	2.5″	26	139
25-200-46	8″	37.10″	29.00″	14.00″	11.50″	1.14″	13.50″	11.75″	8	7.13″	2.5″	35	214
25-250-46	10″	44.82″	34.72″	18.00″	13.00″	1.22″	16.00″	14.25″	12	7.99″	2.5″	37	346
25-300-46	12″	51.84″	39.74″	18.00″	14.00″	1.30″	19.00″	17.00″	12	8.50″	3.0″	44	525

Figure 3-1 Sample shop drawing (figure continues)

Parts List

No.	Part	Material
1	Yoke	Ductile iron, ASTM A536
2	Packing	Asbestos free
3	Bushing	ASTM copper alloy no. C35330, CZ 132
4	O-ring seals	Nitrile rubber, ASTM D2000
5	Valve bonnet	Gray iron, ASTM A126 class B Ductile iron, ASTM A536
6	Bonnet bolt seals	Wax
7	Bonnet gasket	Nitrile rubber, ASTM D2000
8	Bonnet bolts	Zinc coated steel, ASTM A164
9	Stem - A	ASTM copper alloy no. C35330, CZ 132
	Stem - B	ASTM copper alloy no. C99500, low zinc
	Stem - C	Stainless steel 13% chromium
10	Support ring	ASTM copper alloy no. 836, low zinc
11	Pin	Stainless steel
12	Stem holder	ASTM copper alloy no. C35330, CZ 132
13	Wedge, vulcanized	SBR-rubber compound, AWWA C509
14	Wedge body	Gray iron, ASTM A126 class B
15	Nut	ASTM copper alloy no. C35330, CZ 132
16	Pin	Stainless steel
17	Handwheel	Gray iron, ASTM A126 class B
18	Washer	ASTM copper alloy no. C35330, CZ 132
19	Stem nut	ASTM copper alloy no. C35330, CZ 132
20	Stud bolt	Zinc coated steel, ASTM A153
21	Gland follower	Ductile iron, ASTM A536
22	Gland	ASTM copper alloy no. C35330, CZ 132
23	Hexagon nut	Stainless steel, ASTM A320B18
24	Washer	Zinc coated steel
25	Valve body	Gray iron, ASTM A126 class B Ductile iron, ASTM A536

Figure 3-1 Sample shop drawing (continued)

OPERATING INSTRUCTIONS
WATER PRESSURE REDUCING VALVE
FIG. 4500

GENERAL:

The Figure 4500 Valve is designed to maintain a constant delivery pressure regardless of the valve's inlet pressure. The valve is factory tested and set for the required delivery pressure, but field adjustments can be easily made.

INSTALLATION:

The valve is installed with the inlet pressure entering the main valve in accordance with the arrow indication on the drawing or underneath the main valve piston - part #2. Care should be taken to avoid damage to control piping or valve position indicator rod during installation. If the valve is to be repainted, do not paint the indicator rod as it should be smooth and clean to avoid scoring the indicator packing during operation.

OPERATION:

The main valve operates on the differential piston principle such that the area on the underside of the piston is equal to the pipe area while the area on top of the piston is somewhat greater. When equal pressures are applied to both surfaces of the piston, a closing force results that is greater than the opening force, regardless of the line pressure.

To open the main valve, all that is required is to discharge the pressure on top of the piston to the downstream side of the main valve and the inlet pressure acting on the underside of the piston will force the piston open. Trapping pressure on top of the main valve piston will cause the valve to remain in its present throttled position. The downstream pressure should be less than about 80 percent of the inlet pressure for good operation.

SEQUENCE OF OPERATION:

Inlet pressure is constantly being admitted to the top of the main valve piston through a needle valve that is the closing speed control.

A downstream pressure-sensing pilot controls the main valve position by opening, when necessary, and exhausting the pressure above the piston to the valve's downstream side more quickly than water can enter the area above the piston through the needle valve. An increase in the downstream pressure will cause the pilot valve to close slightly. The flow through the needle valve will then exceed the exhaust through the pilot valve, resulting in a downward movement of the piston to a new throttling position.

The needle valve must be in some degree of opening at all times. This valve affects the main closing speed, and generally ¼- and ½-turn opening of the needle valve is adequate. A "hunting tendency of the main valve can usually be dampened by the needle valve. All other hand valves are usually fully open.

ADJUSTMENTS:

Needle Valve	— Closing Speed Control
Hand Valve	— Affects Main Valve Opening Speed If Necessary
Pilot Handwheel	— Turn Counterclockwise to Lower the Downstream Pressure Setting

MAINTENANCE:

When first installed, a slight drip may appear from the air vent tube. This drip should stop when the seals wear themselves in. Should the leakage become excessive, this usually implies that one or both of the seals need replacing.

Failing to maintain a constant pressure into a closed system may indicate the main valve piston seat rubber needs replacing or possibly the pilot seat #2 is damaged.

Erratic operation usually implies the pilot seat or seals are worn or damaged.

If seal replacement seems to occur too frequently on valves 12 in. and larger, the factory can provide different leather seal treatments such as "Thiokol treated leather seals, which are more abrasive resistant. On valves 2½ in. to 8 in. a high hardness rubber may be tried.

Anytime the valve is dismantled, it is recommended that the piston, liner, and seat crown machined surfaces be sanded shiny with a fine emery cloth. This will prolong seal life significantly. A light film of waterproof grease or petroleum jelly should be applied to the leather seals and sliding surfaces.

Periodic inspection of the valve is suggested so that repairs can be anticipated before a malfunction occurs and a system becomes overpressurized.

Figure 3-2 Information from an operations and maintenance manual

INSPECTION AFTER UNLOADING

Inspect valves to verify they are in working order after they are unloaded.

Operation

Large valves should be cycled through one complete opening and closing cycle in the position in which they are to be installed. This process will confirm the following:

- Ease of operation
- Complete travel of the shutoff mechanism direction
- Correct direction of opening
- Required number of turns

Smaller valves may also be cycled, or they can be tested after installation.

Specifications

The initial inspection also should verify compliance with specifications, including the following:

- Size and pressure class (cast in the valve sides so they can easily be confirmed)
- Direction of opening
- Size and shape of operating nut
- Number of turns to open or close
- Type of end connections

Verification of end configuration requires inspection by a person familiar not only with the local specifications and requirements but also with the various end configurations available, such as flanges, mechanical joints, push-on joint ends (various configurations), and combinations of the preceding. ANSI/AWWA C111/A21.11, Rubber-Gasket Joints for Ductile-Iron Pressure Pipe and Fittings, and its references are invaluable in performing this inspection.

Shipping Damage

Inspectors should also check for shipping damage, such as:

- Scoring of the seated surfaces
- Bent stems
- Broken handwheels
- Cracked parts, chipped coatings
- Missing parts and accessories
- Missing lubrication on exterior actuators
- Coating problems, which can be detected with a simple "holiday detector"

After the valves have been inspected, accurate records must be created. Records should include the size and type of valve, time and date of receipt, where the valve is to be stored, comments on inspection and verification, and any comments about the shipment and the way it was received. Chapter 5 discusses record keeping in more detail.

STORAGE

Once records have been made, valves should then be stored in a way that protects them from the environment.

- Valves that are large physical size and weight require the facility to be preplanned to accommodate the valves.
- Metal-seated valves should be stored in the fully closed position to prevent entry of foreign material that could cause damage to the seating surface.
- Resilient-seated valves should be stored in the nearly closed position (rather than fully closed) to avoid unnecessary compression of the resilient material.
- Resilient seats should be protected from sunlight, ozone, and chemical exposure.
- Whenever practical, valves should be stored indoors. If outside storage is required, the operating mechanisms, such as gears, motors, actuators, and cylinders, should be protected from the weather and foreign elements.
- In colder climates, valves should be drained and left slightly open before storage. Failure to do so may result in a cracked valve casting.
- Valves stored outside in cold climates should be stored with the discs in a vertical position. If the discs are in a horizontal flat position, rainwater can accumulate on top of the top disc, seep into the valve body cavity, and then freeze and fracture the casting.
- Electric motor–operated valves should never be stored outdoors because of the sensitivity of electrical contacts to corrosion.

BOLTS

If experience, tests, or observations indicate bolts may be exposed to corrosive environments, the valve may be wrapped or encased with polyethylene sleeving as detailed in ANSI/AWWA C105/A21.5, Polyethylene Encasement for Ductile-Iron Pipe Systems. Alternatively, the valve should use specially coated bolting, such as galvanized, cadmium, or special baked-on resin. Stainless-steel bolts may also be used.

Chapter 4

Installation

Many valve problems and failures can be traced to improper installation, operation, or maintenance procedures, or improper valve application. Valve failure can result in customer disruptions and extensive, costly excavations and repairs.

This chapter presents guidelines for valve installation that usually apply with a focus on buried gate valves installed throughout a water distribution system. Of course, each situation is unique and may require special construction techniques or design.

INSPECTION BEFORE INSTALLATION

At the jobsite, installers should establish a step-by-step routine to ensure that valves are properly inspected before they are installed. The routine should include the following steps:

1. Ensure that the proper valve(s) have been delivered to the jobsite. This includes proper type, size, and pressure rating for the application.
2. Determine the direction of flow in the line to ensure proper installation. Some valves are not bidirectional. Direction of flow is critical when swing-check valves are being installed. Swing-check valves should be mounted with flow in the upward (vertical) flow direction (never the downward flow direction); however, with lever and weight valve styles, the configuration of the weights must assist in closing the valve. Gate valves generally can be installed for flow in either direction.
3. Visually inspect each valve and remove any foreign material in the interior portion of the valve. Foreign material left in the valve can damage internal working parts and may create a contamination problem.
4. Conduct a detailed inspection of the valve similar to the one done on receipt, as discussed in chapter 3.
5. When valves with exposed gearing or operating mechanisms are installed belowground, a vault must be provided to allow pipe clearance and prevent settling on the pipe.

INSTALLATION

Once the preliminary steps have been completed, proper installation generally involves the following steps:

1. Support piping systems and components to minimize alignment issues at the valve connections.
2. Install valves in the closed position to prevent any foreign material from entering the valve.
3. Place the valve on firm footing by providing support in the trench to prevent settling.
4. Install the valve in alignment with the pipe to prohibit excessive strain on the connection to the pipe. Preventing any strain on the connection is critical in flanged installations. Flange fracture can result from unequal tightening of the joints. Figure 4-1 shows the proper technique for installing and tightening the bolts of flanged valves. Important details include the following:
   a. Support and align valves.
   b. Clean dirt and grit particles from all parts.
   c. Insert bolts and nuts and tighten by hand.
   d. Tighten bolts and nuts by using the crossover method shown in Figure 4-1. Tighten the bolts and nuts across the connection from each other until the joint is uniformly tight (do not tighten the bolts and nuts in rotation). This crossover method will load both pipe and valve evenly and eliminate any concentrated stresses.
   e. Ensure that all bonnet bolts are properly tightened to design/manufacturer specifications. Bonnet bolts may have loosened during transportation. Use the same crossover method to verify tightness.
5. Provide a valve box or vault for each valve used in a buried-service application. The valve box should be installed so as not to transmit shock loads or stress to the valve or valve actuator. When valves with exposed gearing or operating mechanisms are installed belowground, a vault must be provided to allow pipe clearance and prevent settling on the pipe.
6. If groundwater or surface water enters the vault, the vault floor should include a recessed drainage area or sump pump to allow for removal of water from the vault. Figures 4-2 through 4-6 show one utility's drawings of its standard valve casing, extension and plug, top and bottom sections, and bonnet. Figure 4-7 shows a cross section of a valve vault.

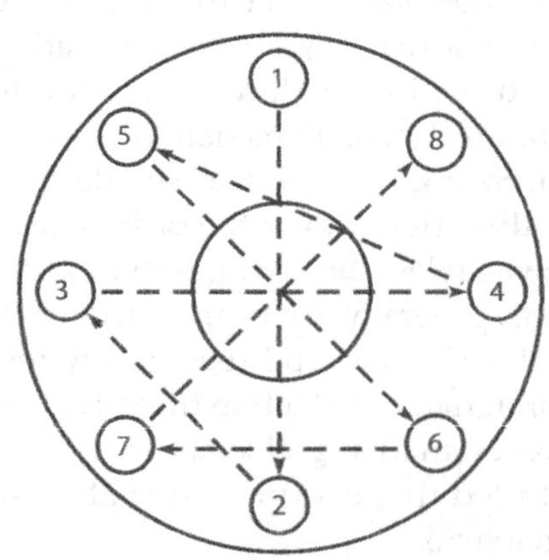

Figure 4-1 Crossover tighten method

7. Center the valve box over the operating nut of the valve. The box cover should be flush with the surface of the finished area. The operating nut should be accessible from the top opening of the vault with a valve key. The valve box should be installed in a manner that will not exert force created from street traffic to the valve.
8. If installing a larger valve that uses smaller bypass valves, install a second valve box over the bypass valve operating nut. Make sure to record the specifics of the bypass valve in the permanent records. The bypass valves play a critical role during shutdowns and startup operations.
9. If the valves must be buried in an unusually deep trench, install a riser on the stem to permit a normal key to be used or make a notation in the valve records (discussed later in this chapter) that a long key will be required for operation.

The following are additional elements for consideration during the installation process related to the corrosion protection of the components:

1. Depending on the corrosiveness of the soil, a passive anode can be inexpensive insurance for aggressive soils, dissimilar metals, stray current, and other conditions that potentially result in external corrosion.
2. Another inexpensive corrosion control method is to poly wrap the pipe and appurtenances, including valves, prior to backfilling in accordance with AWWA standards.

PRESSURE TESTING THE WATER MAIN AND VALVES AFTER INSTALLATION

Where possible and to avoid spending valuable time searching for leaks, crews should wait to backfill valve excavations until pressure tests have been completed. They should perform tests on newly installed piping sections and conduct valve tests at pressure levels higher than the system design pressure. The engineer's or manufacturer's recommended test pressure should be used. Test pressure and allowable leakage for given lengths and types of pipes and joints are given in AWWA standards and manuals.

1. After installation and before pressurization, inspect and tighten all bolts (bonnet, seal-plate, bypass, and end connections) for adequate tightness to prevent leakage. Also, all air must be removed prior to pressure testing the water main. This is generally accomplished by flushing the system. Make sure that all valves that are within the test section (except the end valves) are in the open position.
2. Inspect all tapped and plugged openings of the valve interior for adequate tightness. Proper inspection at this time will minimize the possibility of leaks once the piping system is pressurized.
3. The test pressure should not exceed the rated valve pressure when resilient-seated gate valves or butterfly valves are used to isolate a test section.
4. When metal-seated gate valves are being used as closure pieces for a test, the pressure should not exceed twice the rated valve pressure. Metal-seated valves have an allowable leakage, and this should be determined before the test. For tests at pressures greater than the rated valve pressure, the test setup should include a reduction of the line pressure to the rated valve pressure (independent of the valve) on completion of the test. The valve can then be opened enough to release trapped pressure and equalize it with the line pressure. The valve should not be opened or closed at differential pressures above its rated working pressure.
5. Pressure tests should not be performed on sections of the system that have been connected to the water distribution system, as this could be a source or potential for cross connection and contamination of the potable system. Caution should be taken to ensure proper restraint is provided for all pipe and valve joints especially at dead

ends. The connection to an existing system should be done after the pressure test and biological test are completed.

6. When installation is finished, a complete record of the installation should be made to serve as a reference during maintenance and repair.
7. Once the pressure test is complete, the pipeline section must go through a disinfection test, which needs to be performed according to the engineer requirements, before it is operational.

BEFORE REPLACING AN EXISTING VALVE

Although installation and replacement are very similar, there are a few additional considerations. The routine for a valve replacement should include the following steps:

1. As part of the planning process, several items need to be field verified. Pipe size and material are both important information for planning and ordering replacement parts. It is important to verify the need for restrained joints. Unrestrained pipe has the potential to come undone, quickly flooding the excavation and overwhelming dewatering systems. Field verification can identify other interferences that may need to be addressed before or during the valve replacement.
2. The planning process for replacing a valve needs to include determining how the existing pipe and new valve will be connected. This will ensure the correct parts are on hand. Generally this means either mechanical joint fittings with a short piece of pipe on either side of the new valve, using couplings with a short piece of pipe on either side of the new valve, or replacing a flange valve with another flange valve. Replacing a flange valve with a flange valve will require sufficient space to slide the valve into place for installation.
3. Determine the direction of flow in the line to ensure proper installation. As mentioned previously, some valves are not bidirectional. Direction of flow is critical when swing-check valves are being replaced. Swing-check valves should be mounted with flow in the upward (vertical) flow direction (never the downward flow direction); however, with lever and weight valve styles, the configuration of the weights must assist in closing the valve. Control valves often are designed to flow in one direction.
4. Usually gate valves can be installed to handle flow in either direction.
5. Conduct a detailed inspection of the valve similar to the inspection on receipt, as discussed in chapter 3.

REPLACEMENT VALVE INSTALLATION

Once the preliminary steps have been completed, proper installation generally involves the following steps:

1. Support piping systems and align components to minimize bending at the valve connections.
2. Visually inspect each valve and remove any foreign material in the interior portion of the valve. Foreign material left in the valve can damage internal working parts and may create a contamination problem. Also, clean all pipe sections and coupling as needed. To clean and sanitize the valve, use a 2 percent chlorine solution to clean and disinfect all areas that will come into contact with potable water.
3. Place the valve on firm footing in the trench to prevent settling. The use of a brick or a concrete block may suffice. Do not use wood blocking.
4. Install the valve in alignment with the pipe to prohibit excessive strain on the connection to the pipe. Preventing any strain on the connection is critical in flanged

installations. Flange fracture can result from unequal tightening of the joints. Figure 4-1 shows the proper technique for installing and tightening flanged valves. Details of installation include the following:

a. Support and align valve.
b. Insert bolts and nuts and tighten by hand.
c. Tighten bolts and nuts by using the crossover method shown in Figure 4-1. Tighten the bolts and nuts across the connection from each other until the joint is uniformly tight (do not tighten the bolts and nuts in rotation). This crossover method will load both pipe and valve evenly and eliminate any concentrated stresses. Use the manufacturer's recommended torque specifications.

5. Provide a valve box or vault for each valve used in a buried-service application. The valve box should be installed so as not to transmit shock loads or stress to the valve or valve actuator. When valves with exposed gearing or operating mechanisms are installed belowground, a vault must be provided to allow pipe clearance and prevent settling on the pipe. The valve box and/or vault must be of proper height to allow these components to be flush with the final grade or roadway.

Some additional elements for consideration during the installation process related to the corrosion protection of the components include the following:

1. Depending on the corrosiveness of the soil, a passive anode can be inexpensive insurance for aggressive soils, dissimilar metals, stray current, and other conditions that potentially result in external corrosion.
2. Another inexpensive corrosion control method is to poly wrap the pipe and appurtenances, including valves, prior to backfilling in accordance with AWWA standards.

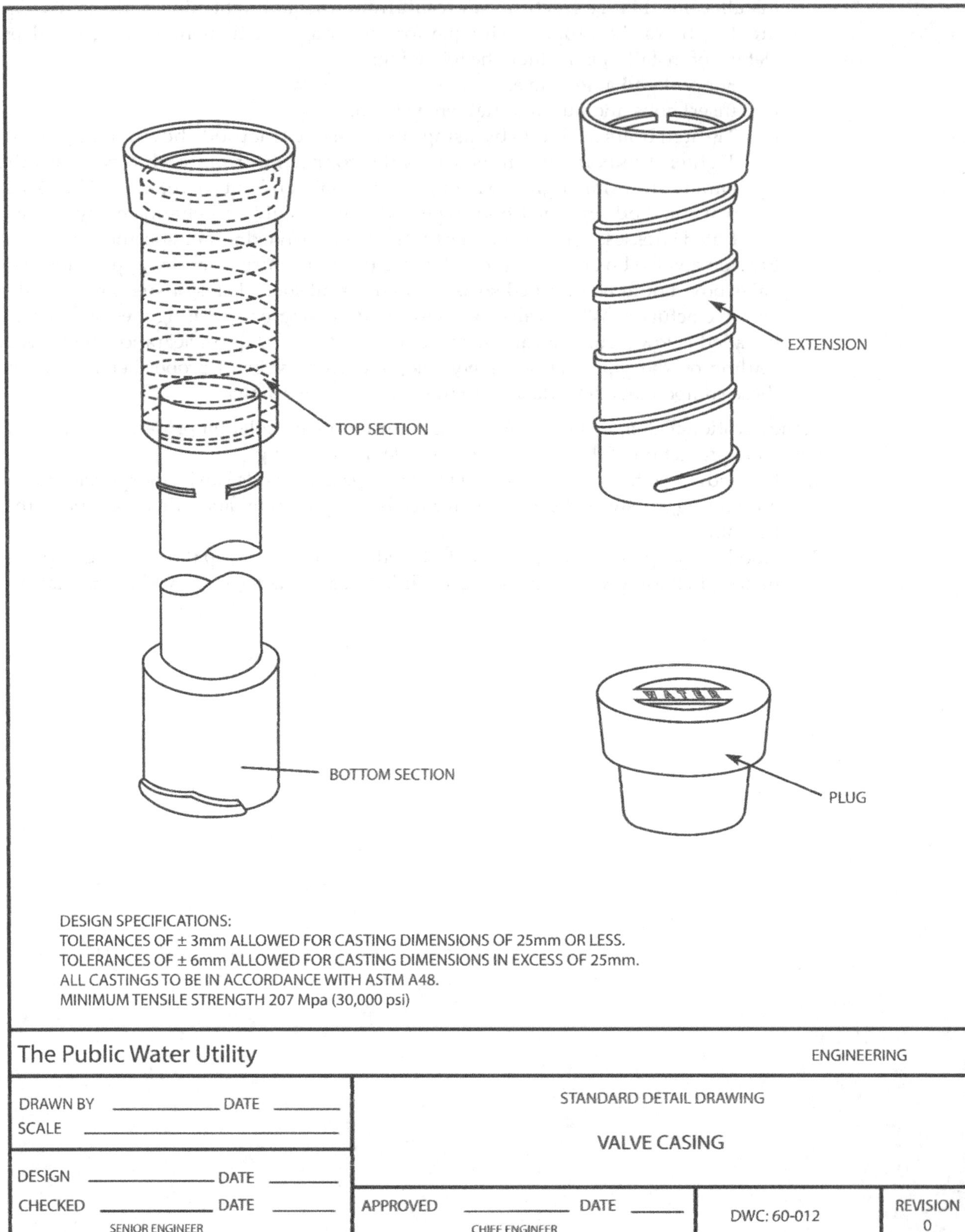

Figure 4-2 Drawing of a typical valve casing

Source: Courtesy Kanwal Oberoi.

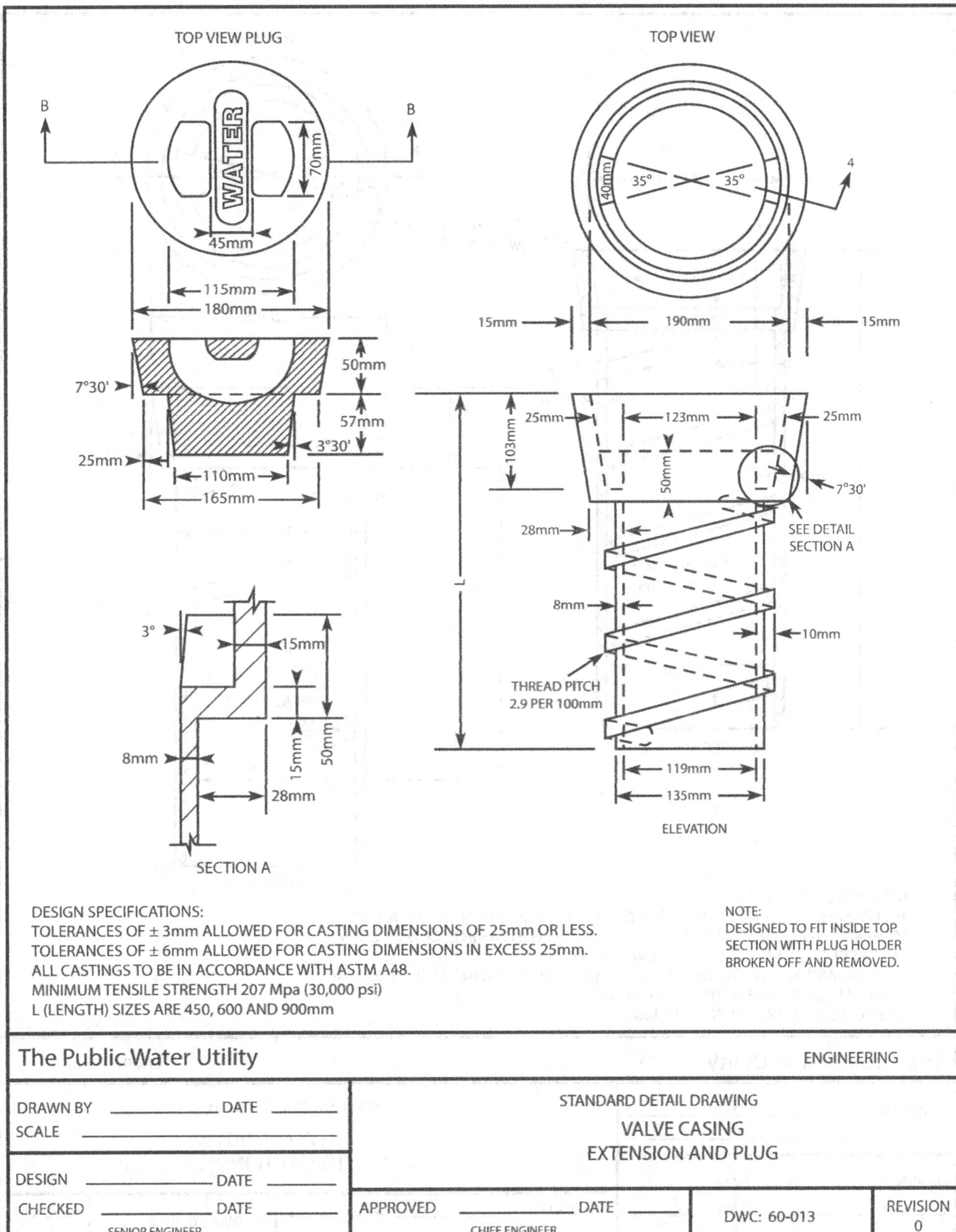

Figure 4-3 Drawing of a typical valve casing extension and plug

Source: Courtesy Kanwal Oberoi.

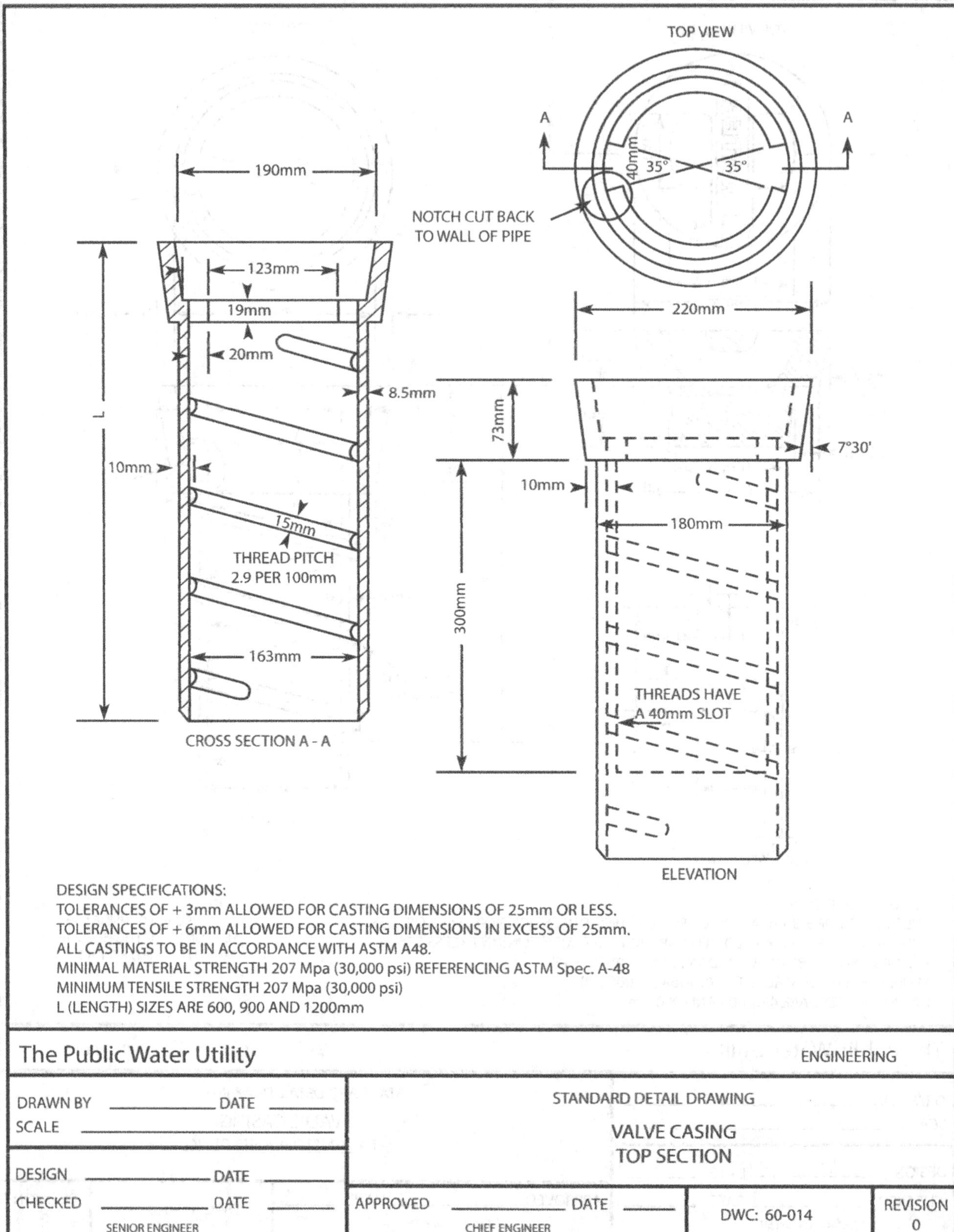

Figure 4-4 Drawing of a typical valve casing top section

Source: Courtesy Kanwal Oberoi.

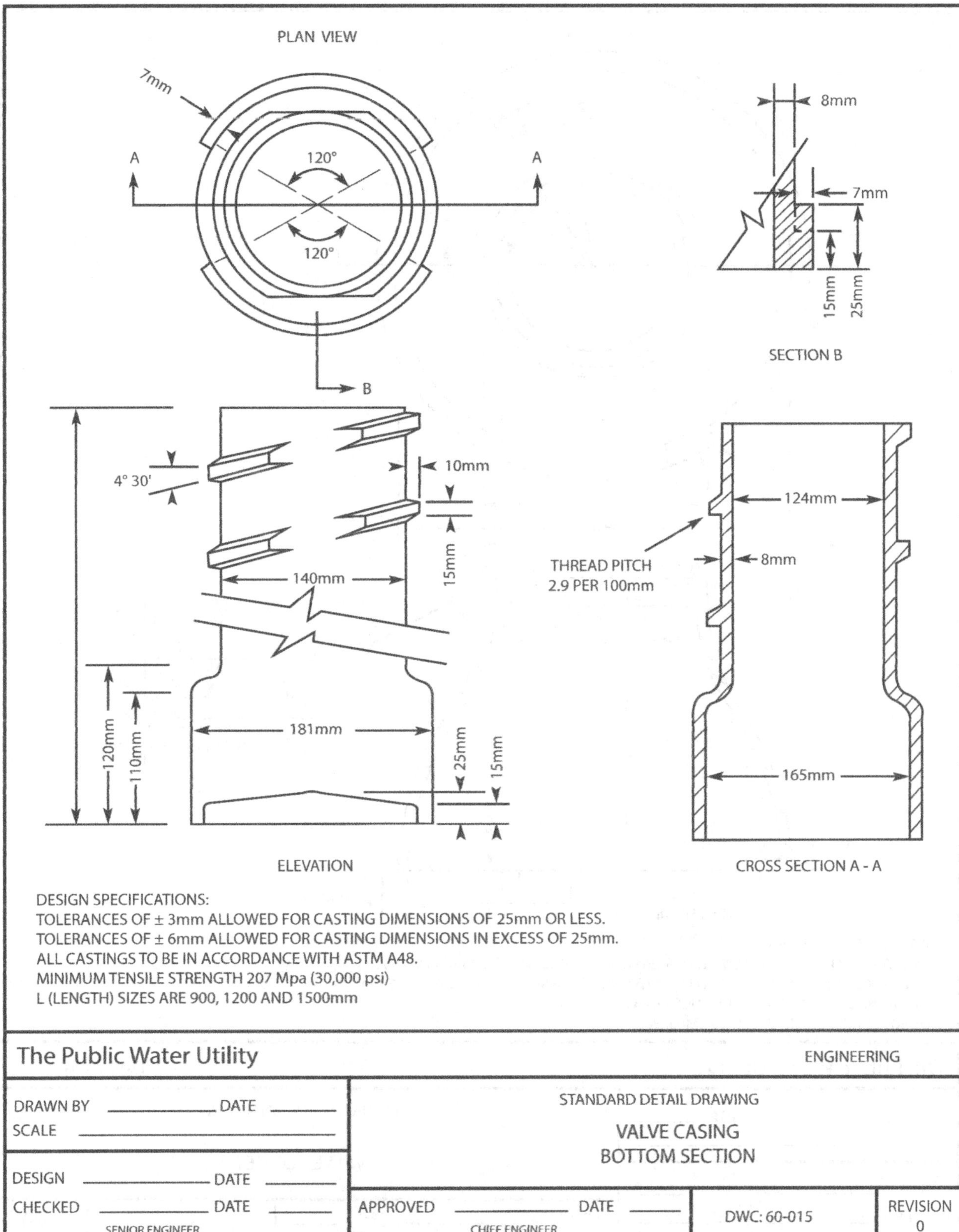

Figure 4-5 Drawing of a typical valve casing bottom section

Source: Courtesy Kanwal Oberoi.

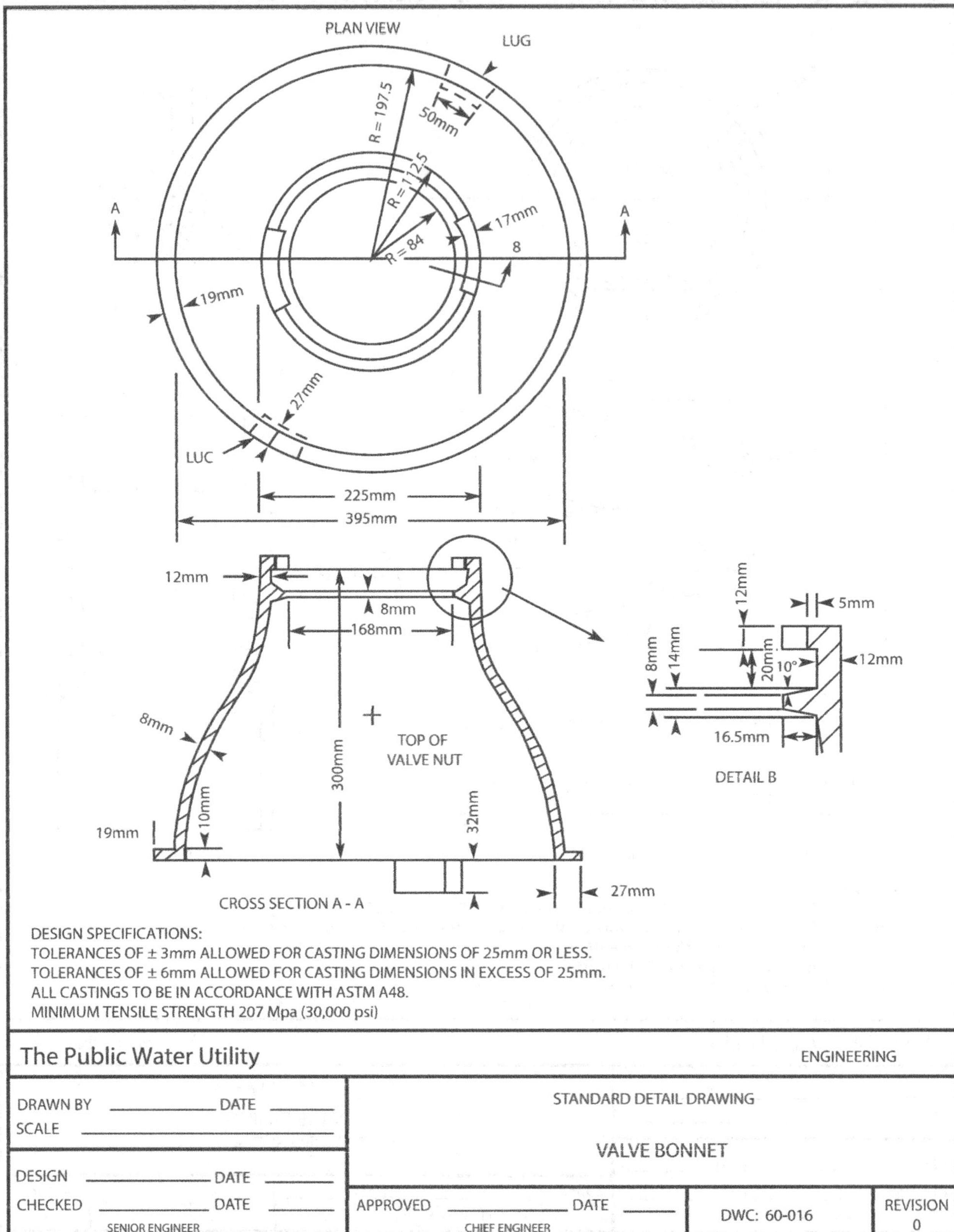

Figure 4-6 Drawing of a typical valve bonnet

Source: Courtesy Kanwal Oberoi.

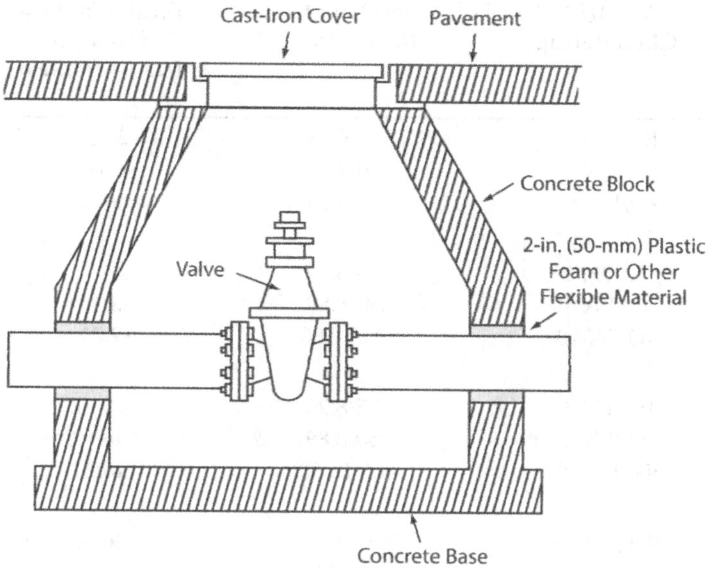

Figure 4-7 Valve vault cross section

Typically, valves placed in a vault as depicted in Figure 4-7 should include a sump area for dewatering purposes. This will facilitate working in a dry environment.

There is great value and importance to torqueing a valve according to the manufacturer's required standard. Correspondence with the manufacturer to get proper torquing specification should be done to facilitate the valve's effective installation and operation. See Table 4-1 for an example of torquing standards.

Table 4-1 Flange bolting specifications*

Valve Size	ANSI/PN Class Rating	Bolt Length (*inches/cm*)	Torque for Low Strength (*ft. lbs./joules*)	Torque for Intermediate Strength (*ft. lbs./joules*)
1-in.	150/PM 16	2.50/6.35	25/31	61/83
DN 25	300/PN 40	3.00/7.62	46/62	122/165
	600/PN 100	3.50/8.89	46/62	122/165
1.5-in.	150/PN 16	2.75/6.99	23/31	61/83
DN 40	300/PN 40	3.50/8.89	82/111	218/296
	600/PN 100	4.25/10.80	82/111	218/296
2-in.	150/PN 16	3.25/8.26	46/62	122/165
DN 50	300/PN 40	3.50/8.89	46/62	122/165
	600/PN 100	4.25/10.80	46/62	122/165
3-in.	150/PN 16	3.50/8.89	46/62	122/165
DN 80	300/PN 40	4.25/10.80	82/111	218/296
	600/PN 100	5.00/12.70	82/111	218/296
4-in.	150/PN 16	3.50/8.89	46/62	122/165
DN 100	300/PN 40	4.50/11.43	82/111	218/296
	600/PN 100	5.75/14.61	132/179	353/479
6-in.	150/PN 16	4.00/10.16	82/111	218/296
DN 150	300/PN 40	4.75/12.07	82/111	218/296
	600/PN 100	6.75/17.15	199/270	531/720
8-in.	150/PN 16	4.25/10.80	82/111	218/296
DN 200	300/PN 40	5.50/13.97	132/179	353/479
	600/PN 100	7.50/19.05	296/401	789/1070
10-in.	150/PN 16	4.50/11.43	132/179	353/479
DN 250	300/PN 40	6.25/15.88	199/270	531/720
	600/PN 100	8.50/21.59	420/570	1119/1517
12-in.	150/PN 16	4.75/12.07	132/179	353/479
DN 300	300/PN 40	6.75/17.15	296/401	789/1070
	600/PN 100	8.75/22.23	420/570	1119/1517

Source: Courtesy of Valtek International.
*Note: Lengths are based upon ANSI Standard B16.5 studs used with raised face bodies

Chapter 5

Valve Asset Management

This chapter focuses on the asset management of valves with specific reference to valve operations and maintenance. Valves are designed to control the flow of water through pipes. Valves are one of the few transmission and distribution system assets that are accessible from the surface. The importance of valve assets continues to increase as buried infrastructure continues to age. If not consistently operated, the probability of effectively opening and closing an older valve decreases over time. Furthermore, properly functioning valves protect the entire distribution system and ultimately the customer. The focus of this chapter is on how to most efficiently and effectively manage valve assets and the risks associated with those assets.

VALVES—DYNAMIC ASSETS

Valves (control assets) are dynamic assets—they are designed to open and close, thereby controlling the flow of water through pipes. Pipes (delivery assets) are static assets—they are designed to be always open and continually allow the flow of water. The dynamic valve assets are point assets, and the static pipe assets are line assets. In water systems, valve assets protect and control static assets. Because valve assets (or at least their operators) are accessible from the surface, valves can be specifically located, assessed, improved, and documented.

Importance of Valve Assets

Asset management is a risk-based management approach in which risk can be characterized as follows:

$$\text{Risk} = \text{Consequence of Failure} \times \text{Probability of Failure}$$

When water systems are newer, there is significantly less probability of failure (breaks, leaks, water quality problems, fire flow reductions, etc.) and the overall risks to the transmission and distribution systems are relatively low. However, with aging underground infrastructure and in some cases inadequate proactive maintenance, the probability of failure increases dramatically. The industry frequently experiences the consequence of inoperable valves. When system valves are broken or inoperable, the section to be isolated becomes much larger, impacting more customers. Additionally, the time required to isolate (and stop the damage) may become much longer because valves are not working. Proactively managed valve systems reduce the consequence of inoperable valves and failures by reducing the area that requires isolation. This reduces the impact to customers while expediting repair time.

Valves—the Physical Asset

There are two primary aspects of the ability for a valve to "control." First, the physical asset must be useable. In order for a physical valve asset to be useable (be able to turn off water flow), the valve must be:

1. *Locatable* (you need to be able to find the right valve in a reasonable period of time even in areas impacted by weather events)
2. *Accessible* (once the correct valve is found, you must be able to access the operating nut on that specific valve)
3. *Operable* (once the correct valve is found and the operating nut is accessible, you must be able to close and open the valve)

If any one (or more) of these critical physical attributes is not achieved, the valve cannot be used to turn on or turn off water. Any failure (locatability, accessibility, or operability of an individual valve) will prevent the valve from fulfilling its purpose. Asset management of valve assets takes into account all three physical facets: locatability, accessibility, and operability.

Valves—the Information Asset

In addition to the physical valve being locatable, accessible, and operable, the information on the specific valve asset must be:

1. *Accurate* (correct and complete information on valve attributes)
2. *Meaningful* (the right attributes are documented and are correct)
3. *Accessible* (the meaningful and accurate information is accessible to the specific person who is going to locate, access, and operate the valve)

If any one (or more) of these critical information attributes is not achieved, the effectiveness of the valve is significantly degraded.

VALVE ASSET MANAGEMENT

Asset management has been described as "a combination of management, financial, economic, engineering, and other practices applied to physical assets with the objective of maximizing the value derived from an asset stock over the whole lifecycle, within the context of delivering appropriate levels of service to customers, communities, and the environment and at an acceptable level of risk" (*International Infrastructure Management Manual* [IPWEA 2006]). Proactive valve asset management is by definition proactively maintaining the level of service required while extending the useful life of a control asset.

Valve Asset Management Components

Asset inventory. The first step to building an asset management program is to compile the known data on the assets. This is usually called an *asset inventory*. It includes items such as the asset indentification (ID); location; GIS (geographic information system) coordinates; size; date of installation; manufacturer; accessibility; valve turns; direction to close; and torque required to turn the valve. Table 5-1, at the end of this chapter, contains a sample asset inventory for valve assets.

Additional elements in prioritizing/managing valve assets are outlined in the following sections and include risk, frequency, geography, and the proactive/planned approach.

Risk. As mentioned previously, risk can be characterized as:

$$\text{Risk} = \text{Consequence of Failure} \times \text{Probability of Failure}$$

As previously discussed, valves control the flow of water and thus protect pipes, tanks, fire hydrants, pump stations, and customers. Much has been discussed in AWWA publications with respect to the criticality of pipe and pipe segments. Like pipes, valves have measurable dimensions of risk and can also be prioritized.

Typical factors that contribute to the *consequence of failure* of valves are:

- The fire protection disruption when shutting down a larger area will be more severe as a result of broken valves
- The area and the number of customers served by the valves that are needed to isolate the section of water main
- The types of customers served by the impacted area can be schools, manufacturers, and hospitals (large number of critical customers)
- The probability of nearby pipe failure (or the probability of a nearby problem such as low pressure or poor water quality)
- The business disruption caused by increasing the area impacted by the shutdown
- The additional traffic disruption caused by a valve not being operable
- The environmental impact of a valve not being usable

Typical factors that contribute to the *probability of failure* of valves are:

- The age of the valve
- The work history or known condition of that valve
- The type of valve, e.g., gate, butterfly, cone, and so on
- The accessibility of the valve (is it in a chamber or buried below the street)
- The manufacturer of the valve
- The utility's history with valves of similar type, size, manufacturer
- The depth of the buried valve
- Corrosive soils or scaling water
- Stray currents

Each water system can have different dimensions and weightings when calculating the risk of valves. In most cases in-line transmission main valves are obviously critical (high risk), but it is not always immediately apparent that a 6-in. sideline isolation valve (off of a transmission main) can also be just as critical. If sufficient detailed system data exists, there are software packages that can help a utility calculate valve criticality. A sample summary criticality matrix (see Table 5-4) showing the probability of failure and consequence of failure for valve assets is included at the end of this chapter.

Frequency. Many utilities use the criticality approach to prioritize asset management activities on valve assets. Typically, valves deemed critical are inspected and exercised on an annual basis, and other less critical valves in the system should be inspected and exercised based on the level of service established by the agency. The intent of valve asset management is to proactively and selectively schedule and perform condition assessment and operations of valve assets in a preplanned manner. Metrics for valve asset management should include written goals and objectives on the process of scheduled asset management activities as well as a tracking mechanism to measure the actual number of assets (by category) per year.

Valve asset management can be a daunting undertaking especially if the utility has not previously systematically performed these activities. Implementing written measurable goals and objectives is an excellent starting point for implementing a sustainable valve asset management program. Each agency should establish its own schedule of valve operation that is consistent with its capacity to perform the work to guarantee the effective operation of the assets. In the end, success not only is a matter of the frequency of inspections but also depends on each agency having a plan in place to regulary inspect its valve assets.

Geography. An alternative to taking a criticality approach, is to methodically work through the entire network from north to south, east to west, tile by tile, or plat by plat. A geographic approach (coupled with criticality) is an excellent sustainable approach to systemwide valve asset management.

Proactive and planned approach when time is available. A preferred approach to valve asset management, and the one demonstrated most successful, is one that provides a written plan, with goals and objectives and dedicated resources directed to proactive asset management activities.

An alternative approach to valve asset management is one in which utility resources (personnel, equipment, processes) are utilized during "down times" to perform proactive asset management activities. The daily demands of system operations and customer service rightly trump proactive activities, and therefore it is difficult for many utilities to drive continuous measurable progress using this approach.

When repair/rehabilitation or replacement is necessary. One of the outcomes of a proactive valve condition assessment and operation program is assets are identified that are evaluated as "unusable" and need some type of repair, rehabilitation, or replacement. Using the criticality analysis, coupled with a cost comparison (of repair versus rehabilitation versus replacement), the utility can also prioritize repairs. If a valve is not usable (and it is open), it acts just like another piece of pipe. Not all valves have the same importance or criticality, and therefore not all repairs/rehabilitations/replacements have the same importance or criticality. How the utility prioritizes and addresses valve work orders is highly dependent on the business processes in place in the agency. Some organizations have crews that operate only, or provide maintenance only, or perform replacements only. Other agencies have crews that provide either a combination of these services or all of them during the course of their assigned work.

Guidelines for Valve Operation and Replacement

Critical elements in valve operation and replacement are outlined in the following sections.

Valve condition assessment and operation. The following is a list of key elements in the assessment and operation of valves:

- Condition assessment includes locating, accessing, and operating the valve, and accurately documenting meaningful information in a location that is accessible.

- Condition assessment and operation of each critical valve and all valves 16 in. and larger should be completed on a regularly scheduled (annual if possible) basis.
- Condition assessment and operation of the majority of valves in a distribution system should be conducted on a three-year to five-year cycle or based on the criteria established by the agency. With historical performance data, the usability decay rate (the rate at which valve assets are rendered unusable due to locatability, accessibility, and/or operability challenges) can be modeled, and this cycle can be further refined to make the best use of asset management investments.
- All gate valves should be physically cycled from full open to close and back open at least once every five years or on a timetable based on the criteria established by the agency.
- When operating a valve, use the lowest torque possible. Do not force the valve.
- If a gate valve is difficult to operate, apply low torque in the closed, then open, then closed direction up to 20 times in an attempt to free up the valve before increasing torque.
- If a gate valve is difficult to operate and has not been operated for an extended period of time, begin with the lowest torque required to turn the valve in the closed direction, moving through 5 to 10 rotations. Reverse for 2 or 3 rotations. Reverse again and rotate 5 to 10 more turns in the closing direction. Repeat this procedure until full closure is attained. Once the valve is fully closed, it should be opened a few turns so that high velocity water flowing under the gates can move the remainder of the sediment in the valve seat. The reason for this cautious approach is that in many gate valves, debris and sediment can build up on the gates, stem, and slides. If this material is compacted while the valve is being closed, the torque required to close the valve continues to build as the material is compacted. If the procedure above is used, the stem and other parts are "scrubbed" by the series of back-and-forth motions, and water in the system can flush the debris that has broken loose away from the stem gate and slides or guides.
- Condition assessment includes examining the condition of the valve box or vault.
- When proactively operating a valve, the valve should be slowly closed, opened, closed, and then reopened. This cycling of the valve should continue until the turn count and the torque required to operate the valve stabilizes.
- Preventive maintenance should be performed as necessary or as suggested by the manufacturer. Some valves (such as butterfly valves that have a seating where a resilient coating meets stainless steel, or valves with actuators isolated from the contents of the line) may need less exercise. The manufacturers' guidelines should be followed.
- During the condition assessment and operation, valves on transmission mains that are associated with the primary source of supply or the only source of supply in a particular area, should only be cycled partially. (In order to not cause an inadvertent outage or hydraulic lock of a gate valve.)
- Caution should be exercised when large valves in critical single-source transmission mains are cycled to the fully closed position.

Valve information. Key information needed about valves may be summarized as follows:

- Condition assessment includes accurately documenting valuable information, including the specific location of the valve asset.

- Some information may only need to be documented once (such as GPS position, type of valve, manufacturer, year of installation, etc.).
- Condition assessment and operation information should be updated in GIS, CMMS (computerized maintenance management system), and/or the utility's EMS (enterprise management system).
- Valve physical data (size, type, manufacturer, date of installation, etc.), location (GPS position), and operational data (torque, turns, close direction, etc.) should be documented digitally. Paper valve cards are useful for documenting historical valve information and this data and future data should be tracked digitally, minimally in an electronic tabular format and preferably in a GIS, CMMS, or EMS. Accurate and up-to-date valve data and information are an integral part of an agency's ability to validate the effectiveness, integrity, and reliability of its system. This information has great value to local fire departments, oversight agencies, and/or regulatory agencies, along with organizations providing ratings for insurance purposes to businesses and homes in the impacted areas.
- The record of the valves should be detailed enough to provide information to reflect the history and reliability.
- Valve records should include, at a minimum, the following information:
  - —Physical characteristics
    - *Size:* 6 in., 8 in., and so on
    - *Type:* double-disk gate, resilient-seated gate, butterfly, cone, check, and so on
    - *Function:* in-line main, hydrant isolation, service line, bypass, division, pressure reducing, and so on
    - *Access:* roadway box, vault, pit, and so on; and lid size
    - *Actuator:* external geared, internal geared, torque reduction device, remote operated, and so on
    - *Cover:* asphalt, concrete, dirt, and so on
  - Location characteristics
    - *GPS coordinates:* submeter or less; from this position the valve can be located via any electronic mapping system or software, and distances from street centerlines can be calculated, as can the street, address, and so on
    - *Asset unique identifier:* ties the unique valve asset to databases and systems
  - Operational characteristics
    - *Direction to close:* close right, close left
    - *Turns:* number of turns required to fully open or close the valve
    - *Position:* open, closed, partially open
    - *Date last operated:* date
    - *Operable:* yes, no
    - *Deficiencies:* detail on any deficiency outstanding that is required to bring the valve assembly back into full usability, such as frozen, requires cleanout, broken stem, rounded operating nut, packing leak, and so on

Additional information, such as manufacturer, year installed, model, depth to operating nut, whether there is a stem extension on the operating nut, and existing work orders, should be documented if the information is available. When documenting valve assets, it is important to realize that much of the information is static (such as the location, size, type, etc.) while a portion of the data is dynamic over time (position, date last operated, operable, etc.).

Valve repair/rehabilitation/replacement. Two observations about valve repair/rehabilitation/replacement are worth noting here:

- Similar to determining the criticality of valves for ongoing asset management (condition assessment and operation), all problems with valves should be prioritized. Repair/rehabilitation/replacement prioritization decisions are dependent not only on the criticality of the valve itself but also on the cost of returning the asset to usability. For example, the highest priority repairs are those where the valve has high criticality and the repair activity is low cost.
- Condition assessment and operation information should be updated in GIS, CMMS, and/or the utility's EMS.

VALVE ASSET MANAGEMENT—SAMPLE RISK SCORING PLAN

The first part of a valve asset management risk scoring plan is the valve asset inventory (see Table 5-1). This inventory is a list of the valves and their respective attributes such as the ID, location, size, year installed, manufacturer (if known), number of turns, torque, and any other attributes the utility wants to record.

The second section of a valve asset management risk scoring plan is the consequence of failure matrix (see Table 5-2). The intent of the matrix is to capture the consequences to the overall water distribution system if the valve is not functional. It is important to remember that while some of the attributes (such as function and redundancy) can be calculated relatively easily, there might be attributes (such as community impact) that require asset managers to use their best judgment.

Table 5-1 Example valve asset inventory

Valve Asset Hierarchy – Sample Plan										
Valve ID	Location	Coordinates		Size	Date installed	Manufacturer	Type	Accessability	# Turns	Original Torque
(#)	(Address)	(X)	(Y)	(in)	(Year)	(Name)	(Name)	(type)	(#)	(lb)
V-1	14 Charles HR Drive	2453.55	5682.46	36	1951	ABC Co.	Gate	vault	120	unknown
V-2	78 AM Boulevard	2765.44	5673.22	12	1984	ABC Co.	Butterfly	buried	39	hand op
V-3	3941 Nuxhall RTHFH	5342.67	2326.43	6	1977	ABC Co.	Gate	buried	21	hand op
V-4	069 TK Heart	7843.23	5732.79	8	1965	ABC Co.	Gate	buried	27	hand op
V-5	55 Davis St	6541.73	4527.35	12	2010	ABC Co.	Ball	buried	39	700
V-6	6 Freel Memorial	5626.78	7825.91	48	1954	ABC Co.	Gate	vault	348	unknown
V-7	7 Boomer Way	7235.67	6713.83	16	1999	ABC Co.	Butterfly	buried	51	hand op

Table 5-2 **Example consequence of failure matrix**

Consequence of Failure Matrix – Valve Asset Management Sample Plan						
Sequence (#)	Valve ID (#)	Function * (1-5)	Redundancy† (1-5)	Critical Needs ‡ (1-5)	Community § (1-5)	Consequence Score ** (0-50)
Weight factor:		30%	40%	15%	15%	
1	V-1	5	5	5	4	48.5
2	V-2	3	3	1	5	30.0
3	V-3	1	1	1	1	10.0
4	V-4	1	1	1	1	10.0
5	V-5	3	3	3	2	28.5
6	V-6	5	5	3	1	41.0
7	V-7	5	3	5	4	40.5

Assign Consequence of failure weight factors for your utility.
* Assign function factors such as Transmission Mains = 5, Cross Conections = 3, Distribution Mains = 1
† Assign redundancy factors: No redundancy = 5; One Redundant Pipe = 3; Lots of Redundancy = 1
‡ Assign factor to represent Critical Needs Consumers: Hospital Affected = 5; Dialysis Unit = 3; Minimal Critical Consumers = 1
§ Assign factors to represent Community Impact: Severe = 5; Moderate = 3; Minimal = 1
** Consequence score is the sum of each weight factor (%) times the criticality factor (1-5). This number is then multiplied by 10.

The third section of the risk scoring plan is the probability of failure matrix (see Table 5-3). Similar to Table 5-2, this matrix is developed by using a combination of known valve attributes and sound engineering judgment to determine the probability that the valve asset will fail within a predetermined timeframe.

As with Table 5-2, the utility will need to determine the appropriate attributes to be scored for each matrix along with the appropriate weight factors to be used. The examples provided herein should be modified to fit the specific needs of the utility.

Table 5-3 **Example probability of failure matrix**

Probability of Failure Matrix – Valve Asset Management Sample Plan						
Sequence (#)	Valve ID (#)	Age * (1-5)	Type † (1-5)	Soil Type ‡ (1-5)	Work History § (1-5)	Probability Score ** (0-50)
Weight factor:		25%	5%	20%	50%	
1	V-1	5	3	0	1	19.0
2	V-2	2	1	3	5	36.5
3	V-3	2	3	3	2	22.5
4	V-4	4	3	1	3	28.5
5	V-5	0	2	5	5	36.0
6	V-6	5	3	0	4	34.0
7	V-7	0	1	5	0	10.5

Assign the weight factor % as appropriate for your utility.
* Assign factors for age such as (60+) = 5, (50-59) = 4; (40-49) = 3; (30-39) = 2; (20-29) = 1; (below 20) = 0
† Assign factors for type such as Cone = 5; Piston = 4; Gate = 3; Ball = 2; Butterfly = 1
‡ Assign factors for soil type such as Highly Corrosive = 5, Moderate = 3; Valves in a Vault = 0
§ Assign factors for work history such as Many Problems = 5, No Problems = 0
** Probability score is the sum of each weight factor (%) times the criticality factor (1-5). This number is then multiplied by 10.

The last step is to multiply the consequence of failure score by the probability of failure score to determine the total risk score for the asset (see Table 5-4).

As stated previously, evaluating buried assets can be very difficult. The utility might elect to place the assets into groups by adjusting the total risk score and performing a more detailed field evaluation on the high risk assets.

Table 5-4 Total risk score

Total Risk Score - Valve Asset Management Sample Plan

Sequence	Valve ID	Location	Size	Date Installed	Consequence Score	Probability Score	Raw Risk Score	Final Risk Score
(#)	(#)	(Address)	(in)	(Year)	(0-50)	(0-50)	(0-2,500)	(1-10)
1	V-1	14 Charles HR Drive	36	1951	48.5	19.0	922	8
2	V-2	78 AM Boulevard	12	1984	30	36.5	1,095	8
3	V-3	3941 Nuxhall RTHFH	6	1977	10	22.5	225	1
4	V-4	069 TK Heart	8	1965	10	28.5	285	1
5	V-5	55 Davis St	12	2010	28.5	36.0	1,026	8
6	V-6	6 Freel Memorial	48	1954	41	34.0	1,394	9
7	V-7	7 Boomer Way	16	1999	40.5	10.5	425	4

NOTE: The Raw Risk Score is dertermined by multiplying the Probability Score by the Consequence Score. This can be used as the final risk score.

However, the Asset Manager might want to develop a weighted Final Risk Score to make comparitive analysis easier. One approach is to divide up the total number of assets into incremental blocks. For example, the top 10% assets all receive a score of 10, the second 10% receive a score of 9. This approach can help to minimize the impact of the subjective scoring system inherent in evaluating buried assets.

This page intentionally blank.

Chapter 6

Distribution Valves Emergency or Shutdown Response Planning

Water distribution system operators are faced with the challenge of efficiently achieving a shutdown to prevent water loss and to restore service to their customers in a timely manner, whether responding to a natural disaster, man-made disaster, water main break, or a scheduled water system shutdown. The implementation of a well-documented maintenance program for distribution valves will greatly enhance an agency's ability to prepare for and quickly recover from all of these situations.

This response can be accomplished by the utilization of the following:

- Accurate valve mapping and valve maintenance records
- Properly maintained valves and appurtenances
- GIS/GPS integration of system infrastructure
- Computerized hydraulic and water quality modeling programs
- A regularly exercised emergency response plan

The identification of critical valves has a greater value than just knowing where these components are located. It provides the ability to isolate sections of the infrastructure using these valves in the repair/replacement process, and this information should be updated regularly. Determining the needed equipment, materials, and personnel to restore service can be more effectively planned with this information. These valves need to be properly mapped, their valve boxes cleaned and uncovered, and exercised regularly. See chapter 5, Valve Asset Management, of this manual for more information on identifying critical valves.

PLANNING A RESPONSE

Regardless of the emergency, the distribution system valving must be relied on to allow access to the system to inspect and repair the mains with minimal water loss and customer outages. As part of the planning process, various scenarios should be considered to better equip the agency for system recovery. Depending on the situation, there are tasks that can be done prior to, during, and after the event to help in the process of bringing the system back into service.

PRIOR TO AN EMERGENCY OR SHUTDOWN EVENT

Prioritizing levels of service is a vital component when planning a response to emergencies. The needs of critical customers such as medical facilities and/or other emergency institutions should be determined to produce an effective response.

The isolation valves can be utilized to separate a portion of the system to reduce the impacted areas in certain disasters. For instance, if the affected area of an emergency or event can be determined prior to its occurrence, specific valves can be identified to be operated to lessen the number of customers who may experience a disruption of service and to reduce water loss from the event. This determination of impact can also be useful during modeling or developing repressurization, chlorinating, and sampling plans. The identified valves need to be labeled and located appropriately in the mapping system or GIS to facilitate their use during an emergency.

When possible, affected customers should be notified of the impending shutdown a minimum of 12 hours before its occurrence as the utility deems appropriate.

DURING THE EVENT

Certain hazards have different impacts on the infrastructure based on their severity, type, and the preparedness of the system. Established goals for the system and particularly the operation of valves help to determine the amount of time it will take when the restoration process begins.

Staff knowledge in the following areas will have a direct impact on the recovery of service:

- Proper operational procedures
- Component and/or system capabilities/complexities
- Redundancy and operational flexibility
- Inventory and location of spare and replacement parts
- The contact information for potential contractors that can assist
- The location and availability of potential required equipment

The operation of valves to isolate the area impacted by the emergency is essential. This isolation could reduce the portion of the system contaminated and reduce water lost during the event, as well as lessening the amount of time necessary to get the components fully operational.

Customers should be notified after service has been restored including any necessary public notifications about the restoration of service and backflow devices, clearing air or discoloration from plumbing, boil advisories, and so on.

AFTER THE EVENT

If the agency has properly planned and practiced its emergency response plan, it will be better prepared to bring the system back online in a timely manner. This preparation, along with the effective maintenance and operation of infrastructure valves, will allow the agency to provide a quality product to the people it serves as well as to reduce the negative impact of any emergency or shutdown. Prior to service restoration, as required by regulators, it is critical that water quality meets regulatory compliance.

An important part of restoring service is making an assessment of the condition of the system. As it relates to valves, this assessment entails the following:

- Determining the number and location of valves impacted
- Evaluating the condition of these valves
- Verifying all valves were reopened
- Verifying that work orders were created to repair/replace the valves that were hard to turn, had rounded hubs, needed to be reset, would not shut off completely, or were free-spinning
- Prioritizing the repair/replacement of valves
- Repairing or replacing the valves if possible before restoring service to customers to prevent future shutdowns and minimize service disruptions
- Utilizing the *necessary* parts and equipment
- Documenting the damage incurred, including loss of service to large areas, to plan to insert needed valves to minimize these out-of-service areas in the future
- Documenting the repair/replacement work completed and restoring any repair parts or other consumables that were expended during the shutdown

POSTINCIDENT REVIEW

After fully restoring the system capacity, conduct a postincident review. Prepare a report documenting the outcome of the review to include:

- A chronology of the event
- A listing of impacted components including water lost
- A condition assessment of the components
- The identification of repair/replacement parts used and whether the current inventory levels were sufficient
- The equipment and supplies utilized
- The number of employees used and their hours worked
- A list of contractors used and an evaluation of each one's response, mobilization time, competence, and job performance
- An assessment of what worked during the process

- The identification of what did not work throughout the event
- A recommendation of what should have been done differently

This review should incorporate the presentation of the information gathered to all appropriate parties, including agency personnel, elected officials, regulatory agency individuals, and the general public. This information can be helpful to the agency in identifying ways to improve efficiency, to operate more safely, and to be appropriately prepared for the next emergency.

The completed incident report should then be stored in an easily retrieved location, ideally within an asset management system, GIS, or CMMS.

References and Additional Readings

American National Standards Institute (ANSI)/AWWA (American Water Works Association). C105/A21.5. Polyethylene Encasement for Ductile-Iron Pipe Systems. Denver, CO: AWWA and ANSI.

ANSI/AWWA. C111/A21.11. Rubber-Gasket Joints for Ductile-Iron Pressure Pipe and Fittings. Denver, CO: AWWA and ANSI.

ANSI/AWWA. C504. Rubber-Seated Butterfly Valves. Denver, CO: AWWA and ANSI.

ANSI/AWWA. C500. Metal-Seated Gate Valves for Water Supply Service. Denver, CO: AWWA and ANSI.

ANSI/AWWA. C507. Ball Valves 6 In. Through 48 In. (150 mm Through 1,500 mm). Denver, CO: AWWA and ANSI.

ANSI/AWWA. C508. Swing-Check Valves for Waterworks Service, 2-In. Through 24-In. (500-mm Through 600-mm) NPS. Denver, CO: AWWA and ANSI.

ANSI/AWWA. C509. Resilient-Seated Gate Valves for Water Supply Service. Denver, CO: AWWA and ANSI.

ANSI/AWWA. C512. Air-Release, Air/Vacuum, and Combination Air Valves for Waterworks Service. Denver, CO: AWWA and ANSI.

ANSI/AWWA. C515. Reduced-Wall, Resilient-Seated Gate Valves for Water Supply Service. Denver, CO: AWWA and ANSI.

ANSI/AWWA. C516. Large-Diameter Rubber-Seated Butterfly Valves, Sizes 78 In. (2,000 mm) and Larger. Denver, CO: AWWA and ANSI.

ANSI/AWWA. C517. Resilient-Seated Cast-Iron Eccentric Plug Valves. Denver, CO: AWWA and ANSI.

ANSI/AWWA. C800.Underground Service Line Valves and Fittings. Denver, CO: AWWA and ANSI.

AWWA. Manual M17, Installation, Field Testing, and Maintenance of Fire Hydrants. Denver, CO: AWWA.

AWWA. Manual M49, Butterfly Valves: Torque, Head Loss, and Cavitation Analysis. Denver, CO: AWWA.

AWWA. Manual M51, Air-Release, Air-Vacuum, and Combination Air Valves. Denver, CO: AWWA.

Institute of Public Works Engineering Australasia (IPWEA). 2006. International Infrastructure Management Manual. 3rd ed. Perth, Australia: IPWEA.

Manufacturing Standardization Society (MSS). Standard Practice 60, Connecting Flange Joint Between Tapping Sleeves and Tapping Valves. Vienna, VA: MSS.

Manufacturing Standardization Society (MSS). Standard Practice 111, Gray-Iron and Ductile-Iron Tapping Sleeves. Vienna, VA: MSS.

Manufacturing Standardization Society (MSS). Standard Practice 113, Connecting Joint Between Tapping Machines and Tapping Valves. Vienna, VA: MSS.

ADDITIONAL READINGS

Lyons, J.L., and C.L. Askland Jr. 1993. *Lyons' Encyclopedia of Valves.* New York: Krieger Publishing Co.

Schweitzer, P.A. 1972. *Handbook of Valves.* New York: Industrial Press Inc.

Skorcz, D. 1983. "Maintaining Distribution System Valves." *Jour. AWWA*, 83(75):556–8.

Skousen, P.L. 2004. *Valve Handbook*. New York: McGraw Hill Inc.

Zappe, R.W. 1991. *Valve Selection Handbook*. Houston, TX: Gulf Publishing Co.

Index

Q

R

S

T

U

V

W

This page intentionally blank.

AWWA Manuals

M1, *Principles of Water Rates, Fees, and Charges,* #30001
M2, *Instrumentation and Control,* #30002
M3, *Safety Management for Water Utilities,* #30003
M4, *Water Fluoridation Principles and Practices,* #30004
M5, *Water Utility Management,* #30005
M6, *Water Meters—Selection, Installation, Testing, and Maintenance,* #30006
M7, *Problem Organisms in Water: Identification and Treatment,* #30007
M9, *Concrete Pressure Pipe,* #30009
M11, *Steel Pipe—A Guide for Design and Installation,* #30011
M12, *Simplified Procedures for Water Examination,* #30012
M14, *Backflow Prevention and Cross-Connection Control: Recommended Practices,* #30014
M17, *Installation, Field Testing, and Maintenance of Fire Hydrants,* #30017
M19, *Emergency Planning for Water Utilities,* #30019
M20, *Water Chlorination/Chloramination Practices and Principles,* #30020
M21, *Groundwater,* #30021
M22, *Sizing Water Service Lines and Meters,* #30022
M23, *PVC Pipe—Design and Installation,* #30023
M24, *Planning for the Distribution of Reclaimed Water,* #30024
M25, *Flexible-Membrane Covers and Linings for Potable-Water Reservoirs,* #30025
M27, *External Corrosion Control for Infrastructure Sustainability,* #30027
M28, *Rehabilitation of Water Mains,* #30028
M29, *Water Utility Capital Financing,* #30029
M30, *Precoat Filtration,* #30030
M31, *Distribution System Requirements for Fire Protection,* #30031
M32, *Computer Modeling of Water Distribution Systems,* #30032
M33, *Flowmeters in Water Supply,* #30033
M36, *Water Audits and Loss Control Programs,* #30036
M37, *Operational Control of Coagulation and Filtration Processes,* #30037
M38, *Electrodialysis and Electrodialysis Reversal,* #30038
M41, *Ductile-Iron Pipe and Fittings,* #30041
M42, *Steel Water-Storage Tanks,* #30042
M44, *Distribution Valves: Selection, Installation, Field Testing, and Maintenance,* #30044
M45, *Fiberglass Pipe Design,* #30045
M46, *Reverse Osmosis and Nanofiltration,* #30046
M47, *Capital Project Delivery,* #30047
M48, *Waterborne Pathogens,* #30048
M49, *Butterfly Valves: Torque, Head Loss, and Cavitation Analysis,* #30049
M50, *Water Resources Planning,* #30050
M51, *Air-Release, Air/Vacuum, and Combination Air Valves,* #30051
M52, *Water Conservation Programs—A Planning Manual,* #30052
M53, *Microfiltration and Ultrafiltration Membranes for Drinking Water,* #30053
M54, *Developing Rates for Small Systems,* #30054
M55, *PE Pipe—Design and Installation,* #30055
M56, *Nitrification Prevention and Control in Drinking Water,* #30056
M57, *Algae: Source to Treatment,* #30057
M58, *Internal Corrosion Control in Water Distribution Systems,* #30058
M60, *Drought Preparedness and Response,* #30060
M61, *Desalination of Seawater,* #30061
M63, *Aquifer Storage and Recovery,* #30063
M65, *On-Site Generation of Hypochlorite,* #30065

This page intentionally blank.